ORGANIC TEA
Cultivation and Marketing

N. Ghosh Hajra
Tea Board of India
Darjeeling Tea Research and Development Centre
Kurseong - 734203, India

International Book Distributing Co.
(Publishing Division)

Published by

INTERNATIONAL BOOK DISTRIBUTING CO.
(Publishing Division)
Chaman Studio Building, 2nd Floor,
Charbagh, Lucknow 226 004 U.P. (INDIA)
Tel. : Off. : 2450004, 2450007, 2459058 Fax : 0522-2458629
E-Mail : ibdco@sancharnet.in

First Edition 2006

ISBN 81-8189-129-7

Composed & Designed at :
Panacea Computers
3rd Floor, Agrawal Sabha Bhawan
Subhash Mohal, Sadar Cantt. Lucknow-226 002
Phone : 0522-2483312, 9335927082
E-mail : prasgupt@rediffmail.com

Printed at:
Salasar Imaging Systems
C-7/5, Lawrence Road Industrial Area
Delhi - 110 035
Tel. : 011-27185653, 9810064311

Dedicated
to
my wife

वासुदेव बनर्जी, भा. प्र. से.
अध्यक्ष
Basudeb Banerjee, IAS
Chairman

14, बी.टी.एम. सरणी (ब्रेबोर्न रोड),
कोलकाता - 700 001
14, B.T.M. Sarani (Brabourne Road)
Kolkata - 700 001

Foreword

Organic farming systems have attracted increasing attention over the last two decades because it has the potential to provide benefits in terms of environmental protection, conservation of non-renewable resources, improved food quality and the reorientation of agriculture towards areas of market demand. The organic tea sector is still a very small part of the tea industry but the number of organic tea producers and the volume of organic tea traded in the world market have recorded high growth over the last few years. This development can be explained by a number of factors viz. awareness among the tea planters of the problem of soil erosion, increased consumer awareness of pesticide residues and heavy metals in conventional teas, and other potential health hazards connected with an intensive system of tea production. Moreover, organic tea has a niche market, where the produce sells at a premium price.

In spite of the considerable media attention there is still very little in the way of published information on what the costs and risk factors involved in farming organically and about the market of organic tea. The present book "Organic Tea— Cultivation and Marketing" provides some ideas on how to address these issues, highlighting questions of standards, equivalence and certification. I hope that all who are associated with the tea industry will find something of interest in this book.

I wish to extend my appreciation as well as to congratulate Dr. N. Ghosh Hajra for his sincere efforts, in the midst of his other engagements, in writing this treatise.

I wish him continued success in publishing other useful publications for the tea sector.

30/11/05

30th November, 2005

B. Banerjee, IAS
Chairman, Tea Board of India

Preface

The gravity of the environmental degradation due to excessive use of chemical fertilizers and pesticides has drawn the attention of the scientists and planners towards finding out ecologically sound, viable and sustainable farming systems for different agro-climatic situations, keeping in view the needs of the present and future generations.

In view of resurgence of interest in alternate agriculture in recent years, the popularity of organic farming is gradually increasing and now organic agriculture is practiced in approximately 130 countries of the world. The area under organic management is growing. In addition, with the increasing degree of health consciousness, the affluent countries are gradually switching their preference towards consumption of organic food products and beverages and are ready to procure these even at much higher price. The market for organic products is, however, continually growing.

The concept of organic farming is gaining ground with the tea planters too. Unlike the market trend for tea in general, demand for organic tea, a value-added product, has been growing rapidly. The organic tea producers could seize the emerging opportunity of this niche world market.

During the last decade I have had the honour to address different aspects of organic tea at various national and international fora and an abridged version of some of those is being compiled in the present volume. Hope, this will find its worth among the tea planters, traders and researchers.

Sincere efforts have been made to ensure the accuracy of the contents of the book. For the general reader a glossary of terms is provided. Numerous boxes provide additional information. Cultivation of organic tea is developing at such a pace that it is inevitable that some of the material contained in this book will have been superseded even by the time it is published. However, I would greatly appreciate the submission of comments or supplemental information on new developments for the future edition. I wish you fruitful reading.

Nikhil Ghosh Hajra

Acknowledgements

I wish to express my gratitude to Mr. B. Banerjee, IAS, Chairman, Tea Board and, Mr. N. K. Das, IAS the then Chairman of Tea Board for providing me the opportunity and inspiration for publication of this book.

The writing of this book was helped considerably by discussions with tea planters, and attendance at national and international meetings.

I am indebted to my colleagues of Darjeeling Tea Research and Development Centre, and friends in the tea industry the world over for their constant support during the preparation of the book.

I would like to acknowledge the help taken in writing this book from various publications and say thanks to the authors for their dedicated work.

I feel obliged and extremely thankful to Mr. Suneel Gomber and his team of International Book Distributing Company, Publishing Division, Lucknow for taking pains during its preparation, printing and publication.

My appreciation also goes to Mr. Partha Dey of Technoworld, Siliguri for computation support at the time of the preparation of the text of this book.

Finally, I am grateful to Mr. Sanjoy Bansal of Sampad Vikash Ltd., Mr. Rajah Banerjee of Makaibari Tea and Trading Co. Pvt. Ltd. ,mfor their inspiration right from the time when this book was contemplated.

This publication does not necessarily reflect the views of the Tea Board of India and in no way anticipates the Board's policy in this area. Its content is the sole responsibility of the author.

Nikhil Ghosh Hajra

Contents

List of Boxes

List of Tables

List of Figures

Chapter 1

Organic Farming in India

The green revolution technologies involving greater use of synthetic agrochemicals such as fertilizers and pesticides with adoption of nutrient responsive, high yielding varieties of crop have boosted the production output per hectare in most of the cases. However, this increase in production has slowed down and in some cases there are indications of decline in growth of crop productivity and production. Priorities in agricultural research are gradually moving from a focus on individual crop performance to total system productivity with due attention on product quality and environment safety. Environmental and health problems associated with agriculture have been increasingly well documented, but it is only recently that the scale of the costs has attracted the attention of planners and scientists. Of late, a growing section of people in the world are questioning the impact of conventional agriculture on environment, economic and social aspects. Consequently, many individuals are seeking, so-called alternative practices that would make agriculture more sustainable and lucrative. Organic farming [1], usage of bio-fertilizer and bio-pesticides, biodynamic farming[2] low input agriculture (http://www.ileia.org/), permaculture[3], sustainable agriculture[4], integrated farming practices (integrated pest management and integrated nutrient management) are some of the practices that are being espoused by proponents both in developed and developing countries. All these practices have evolved as alternatives to chemical use in agriculture.

The roots of organic agriculture developed from differing systems of thought, philosophies of life and agro-political motivations. One thing they all have in common is the desire to form a method of production capable of generating healthful foodstuffs, while limiting any damaging effects on the natural ecosystem. It has in the meantime been scientifically proven beyond doubt that organic farming systems are the most environmentally-friendly, and thus sustainable, agricultural methods. This method of production actively assists in preserving eco-systems and the variety of species, protecting the soil, keeping the water clean and reducing the impact of agriculture on the atmosphere.

Promotion of organic farming

The interest in organic farming in India and other developing countries, however, is growing because it places more reliance on the natural and human resources available, requires less financial input and provides safe food while conserving the environment. Studies to date seem to indicate that organic farming offers comparative advantage in area with less rainfall and relatively low natural and soil fertility levels [5]. Similarly, some areas under tea, coffee, cashew nuts and spices may be easily brought under organic farming with a thrust on exports of organic produce. In other words, rather than promoting organic farming en masse and universally, it would be desirable to carefully delineate areas where either the fertilizer use is nil or nominal (eg. North Eastern States), or are growing crops which could give reasonable yield of high quality produce without the use of agricultural chemicals, ultimately providing better profitability to the farmers, and boost to the export of organic foods.

Among the State Governments, Madhya Pradesh, Nagaland, Sikkim took early lead in promoting organic farming, and Uttaranchal followed the suit and the state has been declared as organic state. Other states like Rajasthan, Karnataka have also initiated steps to give due boost to organic farming in the state. The Karnataka state policy on organic farming was published in 2004. The documents stipulate the policy objectives and a set of strategies for promotion of organic farming in the state. These strategies includes: institutional set up of policy formulation, promotion, production and commercialization of organic products, soil and water conservation, as well as credit facilities, research, and education. At the beginning of 2005, the Department of Agriculture, Karnataka state launched an organic village programme. In this programme, one village of every district will be given support to convert 100 hectares of land to a fully organic system. Out of 27 districts, Krishi Prayoga Pariwara (KPP) is the charge of two (Hulegaru in Srimoga and Kilara in Uttara Kannada district), and Green Foundation of one village. Further, during a national level conference organized at Cochin by UPASI in April 1994, over 400 participants presented their vision of organic agriculture. In April 1995, the Institute for Integrated Rural Development (IIRD) at Aurangabad organized the first networking workshop for Indian members of International Federation of Organic Agriculture Movements (IFOAM). The main recommendations of this workshop included promotion of training and education, development of standards, market development, lobbying and image-building for organic agriculture.

In Central India, the IIRD is involved in training women's groups in organic agriculture, and networks at the national level with NGOs and other groups in organic agriculture. The organization has also set up a school for organic agriculture, the first of its kind in the country. The Sangvi farm and Mr. Save are promoting natural farming techniques in Western India. The Low External Input Sustainable Agriculture (LEISA) network is making headway in different parts of the country by strengthening its network through Agriculture Man Ecology (AME), based in Bangalore. Gloria farms and Auroville in Pondicherry are also contributing towards strengthening the organic agriculture movement in the country. The Society for Organic Agriculture (SOA) of Secunderabad, Andhra Pradesh, is involved in developing marketing strategies for organic products. The Society for Employment Welfare and Agricultural Knowledge (SEWAK) in Nainital is involved in training programmes and documentation of indigenous agricultural practices in Uttaranchal. The India Bio-Organic Tea Association has already been formed to promote organic and biodynamic tea production, and biodynamic practitioners in different parts of India are consolidating their efforts by forming the Association of Biodynamic Movements. The biodynamic movement thus has considerable scope in India.

In August 2003, the Kerala chapter of the Confederation of Indian Industry (CII) organized a workshop on the theme of "Jaivakeralam" at Nelliyampathy. The workshop saw the participation of government officials, NGOs and private sector entities, all focused on evolving a strategy for organic production in Kerala, a state which produces tea, coffee and spices in significant quantities for the export market. In Karnataka, Bangalore-based Transcrop Technologies is promoting organic sugarcane, its agri-chemical division is exploring the possibility of supplying critical inputs to farmers interested in organic sugarcane cultivation. Farmers in and around Chikodi, Belgaun and parts of North Karnataka have evinced interest.

In different parts of India, NGOs and people's organizations are now spearheading the movement for organic agriculture. For example, the Society for Equitable Voluntary Actions (SEVA) in West Bengal, is promoting organic agriculture in 24 Parganas (group of villages or a district subdivision) through training programmes for farmers and through field action programmes. PRAKRUTI (IFOAM members' organisation) at Mahatma Gandhi's Ashram in Sevagram is also playing an important role in mobilizing public opinion for organic agriculture by facilitating marketing of organic agricultural products and organizing organic cotton growers in Western India.

To vitalize the organic agricultural movement, a number of universities and scientists of different research institutes have also started playing key roles. Bidhan Chandra Krishi Viswa Vidyalaya in West Bengal and the University of Agricultural Sciences (UAS), Bangalore have organized national workshops on organic agriculture. UAS has also initiated a correspondence course on organic agriculture so as to reach a large number of farmers. Further, an international conference on "Indian Organic Products - Global Markets" has been organized in New Delhi from 28th November to 2nd December, 2002 by Bio-Dynamic Association of India.

Organic farming does not need costly investments in irrigation, energy and external inputs, but rather organic farming policies have the potential to improve local food security, essentially in marginal areas [5]. Further, organic farming in the Indian context can be significant [7] in two distinct ways :

- *To increase the efficiency and sustainability of production*: Organic farming can help to reduce production costs (especially where labour is cheap compared to input costs) and to increase or stabilize yields on marginal soils. This is especially relevant for smallholders in marginal areas where green revolution agriculture has lead to a depletion of soil fertility and to high debts because of increase in input costs.
- *To increase product value*: In areas where farmers have access to established organic markets within the country or abroad, products can achieve a higher price compared to the conventional market. Especially in the trend of decreasing prices for agricultural products, this can be an important way to stabilize or even increase incomes.

Box 1.1

Philosophy of Organic Farming

Organic farming can be seen as an approach to agriculture where the aim is to create integrated, humane, environmentally and economically sustainable agricultural production systems. The 'organic' is best thought of as referring not to the type of inputs used, but to the concept of the farm as an organism, in which all components - the organic matter, soil minerals, insects, plants, animals, microorganisms and humans - interact to create a coherent, self-regulating and stable whole. Reliance on external inputs, whether chemical or organic, is reduced as far as possible. In many European countries organic agriculture is known as ecological or biological agriculture, reflecting the reliance on ecosystem management rather than external inputs.

Source: Ref. 8.

The concept of organic farming

India has traditionally practiced organic agriculture. Several forms of organic farming are being successfully practiced in diverse climate, particularly in rainfed, tribal, mountains and hilly areas of India. Further, the importance of organic manures in agriculture is known since ancient times, and finds mention in ancient Hindu religious scriptures (Rig Veda 1, 161, 10, 2500-1500 BC; Atharva Veda II 8.3). The Holy Quran also mentions that at least one-third of what you take out from soils must be returned to it, implying recycling of post-harvest residues. However, the organic movement in India owes its origin primarily to the work of Sir Albert Howard, who believed that a shift from nature's methods of crop production to adoption of newer methods led to the loss of soil fertility [9]. It is this retrogressive thinking which possibly sowed the seeds of organic movement in India placing great emphasis on the use of compost and other organic sources of plant nutrients to the total exclusion of chemical fertilizers[10].

The concept of organic farming has been perceived differently by the farmers of the developing and, of the least developed countries. To most of the farmers in countries like India, it implies the use of the organic manures and natural methods of plant protection instead of synthetic fertilizers and pesticides. It is also regarded by some as a farming involving the combined use of fertilizers and organic manures as nutrient sources as well as of chemicals and natural inputs for plant protection. In all these instances, the concept has been understood only partially[1].

Definition of organic farming

There is no single definition for organic farming as the term refers to a movement rather than to a single policy. In fact, organic agriculture is based on holistic management systems which promote and enhance agro-ecosystem health, including biodiversity, biological cycles, and soil biological activity. It emphasises the use of management practices in preference to the use of off-farm inputs, taking into account that regional conditions require locally-adapted systems. In the context of the Codex Alimentarius Guidelines, "organic" is a labelling term which denotes products that have been produced in accordance with organic production standards and are certified by a duly constituted certification body or authority. Other terms can be used, such as "biological" and "ecological", depending on the country. "biodynamic" is also covered under this term, but it includes additional practices [11].

The International Federation of Organic Agricultural Movements (IFOAM) has defined the objectives of organic farming as to:

- produce food and beverages of high nutritional quality in sufficient quantity;
- work with natural systems rather than seeking to dominate them;
- encourage and enhance biological cycles within the farming system, involving micro-organisms; soil flora and fauna, plants and animals;
- maintain and increase the long-term fertility of soils;
- use as far as possible renewable on-farm resources in locally organized agricultural systems;
- work as much as possible within a closed system with regard to organic matter and nutrient elements;
- give all livestock conditions of life that allow them to perform all aspects of their innate behaviour;
- avoid all forms of pollution that may result from agricultural techniques;
- maintain the genetic diversity of the agricultural system and its surroundings, including the protection of plant and wildlife habitats;
- allow agricultural producers an adequate return and satisfaction from their work including a safe working environment;
- consider the wider social and ecological impact of the farming system.

These principles provide the basis for day-to-day farming practices to the organic farmers of the world. The United States Department of Agriculture has framed a handy definition of organic farming which, although it misses out some important aspects [12], provides a description of the key practices:

Organic farming is a production system which avoids or largely excludes the use of synthetically compounded fertilizers, pesticides, growth regulators and livestock feed additives. To the maximum extent feasible, organic farming systems rely on crop rotations, crop residues, animal manures, legumes, green manures, off-farm organic wastes, and aspects of biological pest control to maintain soil productivity and tilth, to supply plant nutrients and to control insects, weeds and other pests.

The concept of the soil as a living system...that develops...the activity of beneficial organisms...is central to this definition [12].

Land area

Worldwide, at least 130 countries produce certified organic foods and beverages in commercial quantities, 50 percent are from developing countries [13]. According to the latest survey more than 26 million hectares are managed organically world-wide [14]. The countries with the largest areas of organic farmland are: Australia (about 11.3 million hectares), Argentina (almost 2.8 million hectares), Italy (almost 1 million hectares) and USA (0.9 million hectares). In Europe, more than 6.3 million hectares are under organic management. In total, Oceania holds 43 percent of the world's organic land, followed by Europe (23.8 percent) (Fig. 1.1). In most Asian countries the area under organic agriculture is still very low, but it is increasing rapidly. Among the more significant countries producing organic products are China, India, Indonesia and Japan. The total organic area in Asia is about 736,000 hectares.

Estimating area under organic agriculture in India is a difficult task as there is no central agency that collects and compiles this information. Different agencies have estimated the area under organic

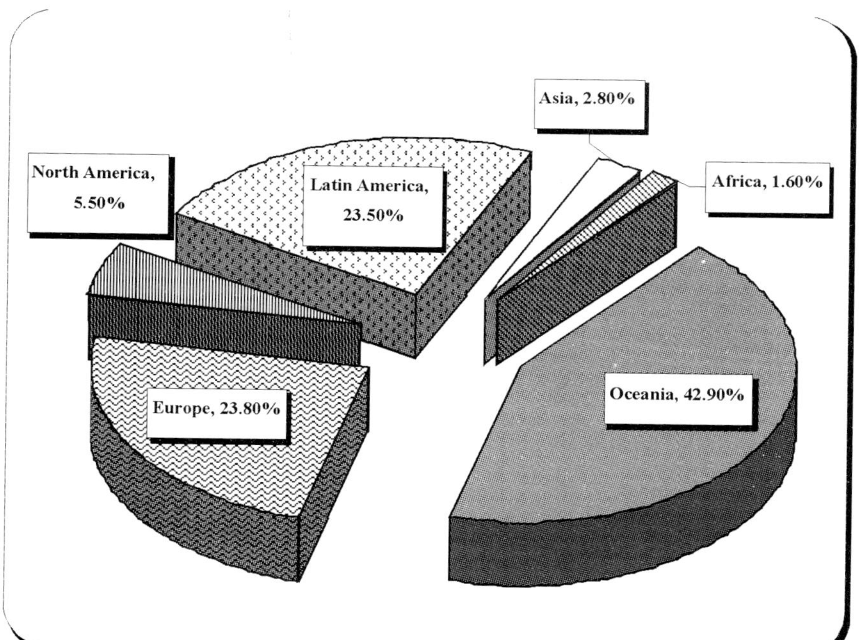

Fig. 1.1. Total area under organic management – Share for each continent (Source: Ref. 14)

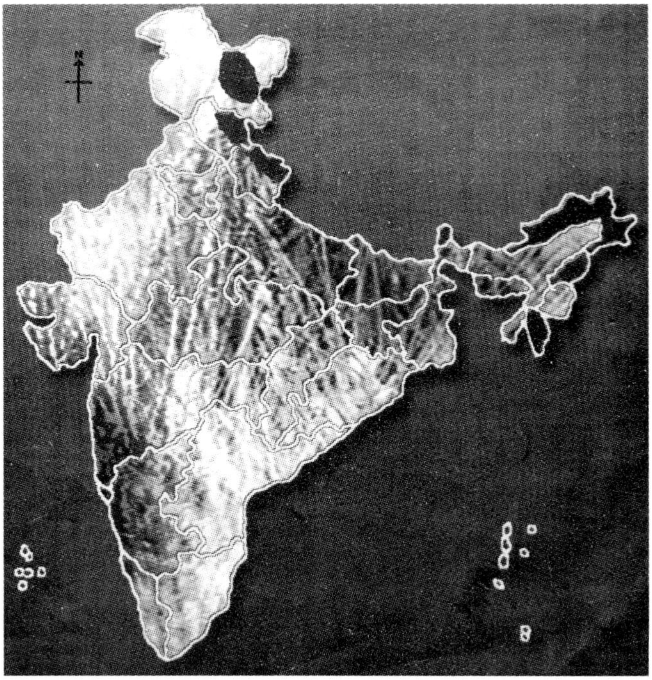

Fig. 1.2. Important existing organic farming regions of India (Source: Fertilizer Association of India)

agriculture differently for instance the study undertaken by FiBL and ACNielsen ORG-MARG the area under organic agriculture to be 2,775 hectares (0.0015% of gross cultivated area in India) [7]. But other estimation undertaken by SOEL-Survey shows that Indian agriculture as a whole accounts for about 180 million hectares, the total land area under organic management (Fig. 1.2) is 41,000 ha and in percent of total agricultural area is 0.03. Around 5,147 farms in India have been certified as organic farms [14].

Indian organic products

The production of organic crops in India is around 14,000 tons [7]. The main organically produced product groups which are traded internationally from India are: fresh fruit and vegetables, dried fruits and nuts, processed fruit and vegetables, coffee, tea, cocoa, spices and herbs, oil crops and derived products, sweeteners grains, dried leguminous vegetables, meat, dairy products, eggs, alcoholic beverages, processed food/food preparations (Table 1.1). In addition, some non-food products, e.g. feeding stuff, seed grains and cotton, should also be mentioned. Tea and rice contributes around 24 percent each, fruits and vegetables combine makes 17 percent of this total production (Fig. 1.3). India is best known as an exporter of organic tea and also has great export potential for many other products.

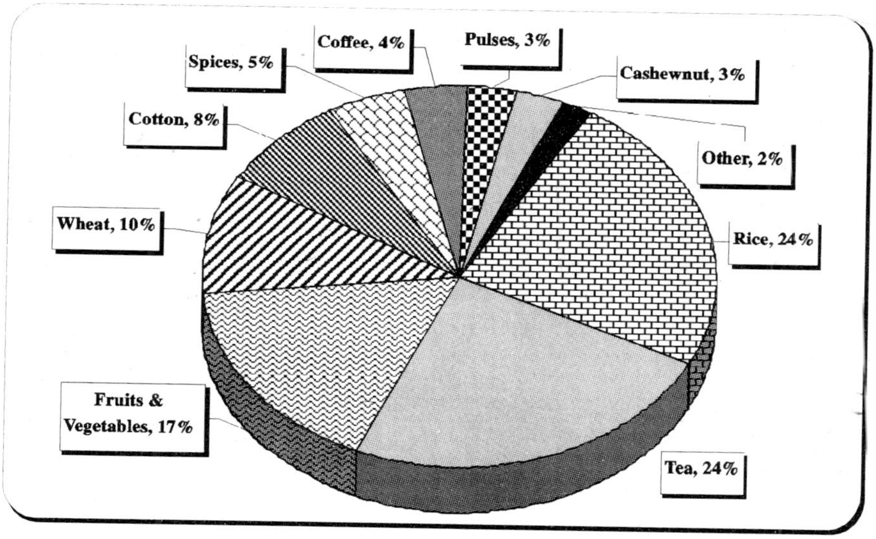

Fig. 1.3. Percent contribution of different crops grown organically in comparison to total organic production in India (Source: ACNielsen Org-Marg, 2002).

Table 1.1. Major products produced by organic farming in India.

Type	Products
Commodity	Tea, coffee, rice ,wheat
Spices	Cardamom, black pepper, white pepper, ginger, turmeric, vanilla, mustard, tamarind, clove, cinnamon, nutmeg, mace, chili
Pulses	Red gram, black gram
Fruits	Mango, banana, pineapple, passion fruit, sugarcane, orange, cashew nut, walnut
Vegetables	Okra, brinjal, garlic, onion, tomato, potato
Oil seeds	Sesame, castor, sunflower
Others	Cotton, herbal extracts

Source: Ref. 7

Advantages for Organic Production in India

Diverse agro-climatic regions across the country that provides environment for wide range of crops that can cater to different market demands. Some comparative advantages for organic production in India [15] are:

1. Growing export market for organically-produced crop.
2. India is strong in high quality production of certain crops like tea, some spices, rice specialties, ayurvedic herbs etc.
3. India has a rich heritage of agricultural traditions that are suitable for designing organic production systems. Sophisticated crop rotation or mixed cropping patterns, for example the famous agro-forestry systems of the Western Ghats, facilitate the management of pests, diseases and nutrient recycling. Botanical preparations, some of which originate from the ancient Veda scripts, provide a rich source for locally-adapted pest and disease management techniques. The widespread cultivation of legume crops facilitates the supply of biologically fixed nitrogen.
4. In several regions of India agriculture is not very intensive as regards the use of agro-chemicals. Especially in mountain areas and tribal areas, use of agro-chemicals is rather low, which facilitates conversion to organic production. On these marginal soils, organic production techniques have proved to achieve comparable or in some cases (especially in the humid tropics) even higher yields than conventional farming.
5. Availability of comparatively cheap labour for labour-intensive organic agriculture.
6. Increasing awareness and health consciousness especially among certain sectors of domestic consumers.
7. The Non-Governmental Organization (NGO) sector in India is very strong and active promoters of organic farming provide training, extension services information and marketing services to farming communities in different agroclimatic regions [16].
8. Enhanced Government attention and support for organic agriculture through various policy initiations and action programmes.

Government initiatives to promote organic farming

At present, India is in the nascent stage of organic farming. To promote the organic agriculture in India the Government of India, has taken some initiatives in the recent past. In order to encourage the organic agriculture movement in India, the Government has launched the National Programme for Organic Production (NPOP) on 8 May, 2000 with the following declared objectives [17]:

- Declaring standards for organic production;
- Recognizing specific bodies for preparing an approved package of practices for specified products and for approving certification programmes;
- Allowing the recognized bodies to accredit agencies to inspect and certify those products;
- Seeking recognition from, and according reciprocal recognition to, standards of other nations and trading blocs;
- Instituting a logo and prescribing its award by accrediting bodies to products that qualify for bearing the Indian Organic Label.

Under the NPOP, National Standards (available at www.apeda.com/organic/ quality.html) were published in 2000. These are based on IFOAM's Basic Standards of Organic Agriculture and Food Processing, with suitable modifications, taking into account the agricultural and climatic conditions prevailing in India. A National Steering Committee functioning under the Ministry of Commerce is monitoring the programme of NPOP has been established with representatives from the Government and trade to function as the apex advisory body for assisting the Government in shaping the growth and development of organic agriculture. Agricultural and Processed Food Products Export Development Authority (APEDA) under the Ministry of Commerce is the nodal agency to promote the Indian organic agriculture and its export opportunities.

As a part of this programme, a National Organic Logo, "Indian Organic", communicating the authenticity and the origin of the product, will be granted on the basis of compliance with the National Standards of Organic Products (NSOP) (http://www apeda.com/organic.htm).Some of the other efforts towards promotion of organic exports include attempts to collaborate with all the major organic importing countries. Towards this, APEDA is deliberating with European Union for inclusion of India in the list of third countries under article 11 of the EU regulation No 2092/91 so that Indian National Programme for Organic Production gets the required recognition under the EU regulations[18].

In the context of gradual expansion of organic farming, a proper regulatory mechanism for quality assurance is needed. Regulatory mechanism is nothing but a controlling system which will be guided by specific standard, a principal of properties, honesty and integrity or a level of excellence of quality. Adoption of organic agriculture necessarily involves a sequence of steps that need to be followed by the growers and verified by certification and inspection agency. The 'regulatory' word refers to a mandatory, affordable, accessible and responsible programme of organic certification. Its primary aim is to protect the interest of consumers.

Guidelines for regulatory mechanism

During July, 2002, the Government of India circulated guidelines on promotion of organic farming. As per para (3) of this guideline, a sequence of steps for regulatory mechanism on porganic farming has been indicated and this includes:

- formation of organic farmer group;
- registration of farmers group with district authorities;
- documentation of individual farms/farmers records;
- service provider;
- accreditation agency;
- certification and inspection agencies;
- formulation of National Standards of organic farming;
- conversion periods;
- crop production;
- manurial policy;
- pest, disease and weed management;
- monitoring along with list of permitted/non-permitted inputs/ process.

Accreditation agencies

The mandate of the accreditation agencies includes, among others, the prescription of a package of practices for organic products within their jurisdiction and accreditation of certification agencies. The inspection and certification charges to be levied by the certification agencies are to be fixed by the accreditation agencies. On the basis of the mandate given by the NPOP, the accreditation agencies have issued detailed regulations covering, among others, standards, packages of practices and procedures for accreditation of certification agencies for their mandated products[6]. Presently, the six accreditation agencies approved by the Ministry of Commerce, Government of India are authorized:

• Tea Board • Coffee Board • Spices Board • Coconut Development Board • Directorate of Cashew and Cocoa Board • Agricultural and Processed Food Products Export Development Authority (APEDA).

Certification agencies

The accredited certifying agencies in India are:

Inspection and Certification agencies of India accredited under the NPOP*

Ecocert SA (India Branch Office)
Sector-3, S-6/3 & 4, Gut No. 102
Hindustan Awas Ltd., Walmi-Waluj Road
Nakshatrawadi, Aurangabad – 431 002, Maharashtra
Phone No.: 0240-2377120, 2376949
Fax No.: 0240-2376866
Email: ecocert@sancharnet.in

BVQI (India) Pvt. Ltd.
Marwah Centre, 6th Floor, Opp. Ansa Industrial Estate, Krishanlal Marwah Marg, Off Saki-Vihar Road, Andheri (East), Mumbai-400 072, Maharashtra
Phone No.: 022-56956300, 56956311
Fax No. 022-56956302 / 10
Email: scsinfo@in.bureauveritas

IMO Control Private Limited
No. 1314, Double Road, Indiranagar 2nd Stage
Bangalore-560 038, Karnataka
Phone No.: 080-25285883, 2520 1546
Fax: 080-25272185, Email: imoind@vsnl.com

Indian Organic Certification Agency (INDOCERT)
Thottumugham, P.O. Aluva-683 105
Cochin, Kerala
Telefax:0484-2630908-09/2620943
Email: Mathew.Sebastian@indocert.org

Lacon Quality Certification Pvt. Ltd
Chenathra, Theepany, Thiruvalla - 689 101, Kerala
Telefax: 0469 2606447
Email: laconindia@sancharnet.in

SGS India Pvt. Ltd.
250 Udyog Vihar, Phase – IV, Gurgaon – 122 015
Haryana
Phone No.: 95124-2399990 to 98
Fax No.: 95124-2399764
Email: sudarshan_sharma@sgs.com

Skal International (India)
A Division of CU Inspections India Pvt. Ltd.
No. 191, 1st Main Road, Mahalaxmi Layout
Bangalore – 560 086, Karnataka
Phone No.: 080-23491928, 56966507
Fax no.: 080-23491935
Email: skalindia@eth.net

International Resources for Fairer Trade
Sona Udyog (Industrial Estate)
Unit No. 7, Parsi Pandhayat Road, Andheri (E)
Mumbai – 400 072, Maharashtra
Phone No.: Tel: 022-28352811, 28235246 ext. 22
Fax – 022-823-5245
Email: irft@vsnl.com

Natural Organic Certification Association
11-B, Kulkarni Bungalow, 5th Lane, Shikshak Nagar
Kothrud, Pune-411 038, Maharashtra
Phone No.: 020-5621-8063, Fax: 020-2539-0096
Email: contact@nocaindia.com

OneCert Asia Agri Certification Private Limited
Agrasen Farm, Vatika Road, Off Tonk,
Jaipur-303 905, Rajasthan
Telefax No:0141-2720202
Email: info@onecertasia.in

Uttaranchal State Organic Certification Agency
(USOCA), 12/II Vasant Vihar, Dehradun-248 006
Uttaranchal
Phone No.: 0135-2760861
Fax: 0135-2760734
Email: uss_opca@rediffmail.com

* May visit http://www.apeda.com/organic/ for details of annual charges.

Certification

Certifying agencies play an important role for producers in facilitating access to organic markets, especially on the export level. To regulate the export of certified organic products, the Director

General of Foreign Trade, Government of India has issued a public notice according to which no certified organic products may be exported unless they are certified by an inspection and certifying agency duly accredited by one of the accreditation agencies designated by the Government of India.

Cost of inspection and certification

An example of the fee structure of a certification body operating in India is presented in Table 1.2.

Table 1.2. Cost of inspection and certification.

Category	Details	Fees (Rs)
Small farmers and co-operatives	Travel and inspection Report preparation Certification	12000/day 5000 flat fee 5000/certificate
Estate manufacturers and Exporters	Travel and inspection Report preparation Certification	19200/day 5000 flat fee 5000/certificate
Large and medium-sized processors	Travel and inspection Report preparation Certification	16800/day 5000 flat fee 5000/certificate

Source: ACNielsen ORG-MARG, 2002 (Field survey and the publication - Organic and Biodynamic farming, Govt. of India, Planning Commission)

Over the past few years, many international certifiers have opened branch offices in India and operated by the Indians. Thus, the cost for certification has come down considerably. At the same time, local certification bodies started to emerge, partially with the aim of further reducing certification costs.

The basic infrastructure for regulating the growth of organic agriculture in India has thus been established. However, much remains to be done. These includes obtaining recognition for the Indian standards by international standards organizations, including those in important markets; involving producers to ensure that the standards adequately reflect field situations; and accrediting credible certification agencies [6].

Export market

The Organic agricultural export market is one of the major drivers of greening of agriculture in India. Around 11,925 tons of organic products are exported from India that makes around 85 percent of total organic crop production. The major export market for Indian producers are Australia, Belgium, Canada, France, Germany, Italy, Japan, Netherlands, Sweden, Singapore, South Africa, Saudi Arabia, UAE, UK, and USA. In general, German and Dutch traders are importing mainly spices, fruits, tea, and nuts. Traders from the Netherlands are interested to import of sesame. Swiss traders are mostly importing tea, cotton and organic spices etc. Estimated quantity of various products that are exported from India in 2002 is shown in Table 1.3. This shows that around 3000 tons of tea was exported and in quantity term it was the highest, next major exports are rice (2500 tons), fruits & vegetables (1800 tons), cotton (1200 tons) and wheat (1150 tons)[7]. As seen from the global market growth trends for organic foods, there are enormous potentials for India to explore. Some of the prerequisites for exploiting this potential include:

a. farmers' capacity to produce organically the agricultural products which have global market; and

b. prior experience of exporters and traders in exporting agricultural commodities to these markets.

The attractiveness of organic market gets enhanced also because of the price premium. Price premium for various organic produces vary in different countries depending upon the distribution channels and market conditions. The details of price premiums and international market for organic products will be dealt with in chapter 10 and 11 respectively.

Table 1.3. Major organic products exported from India and their sales in 2002.

Product	Sales (tons)
Tea	3000
Coffee	550
Spices	700
Rice	2500
Wheat	1150
Pulses	300
Oil Seeds	100
Fruits & Vegetables	1800
Cashew Nut	375
Cotton	1200
Herbal Products	250
Total	11,925

Source: ACNielsen Org-Marg, 2002

Kolanu and coworker [18] made an attempt to develop a matrix (Table 1.4) by depicting the current conventional tea market indicates the capabilities of India in exporting tea to different countries and similarly depicting the existence of organic market for tea in these countries reveal the existence of opportunities for exporting organic tea. The matrix reveals that India has demonstrated capabilities of exporting tea to countries like USA, U.K, Germany, Japan, France, South Africa, CIS Countries, Poland, Netherlands and Italy. It also shows that there is a demand for organic tea in most of these countries, which attract price premiums ranging from 10 to 100 percent. This showcases a window of opportunity that is yet to be exploited to its maximum potentials by Indian organic tea exporters and producers.

Domestic market

The domestic market for organic products in India is not yet developed as the export market. The products available in the domestic market in organic quality are rice, wheat, tea, coffee, pulses, fruits and vegetables (ACNielsen ORG-MARG Survey, 2002). To a large extent this sale is also based on individual initiative of the farmers, NGOs (Non Governmental Organization) and some

Table. 1.4. Existing conventional export market and prospective market for Indian organic tea.

Country	Existing conventional Export market	Propective market for organic tea
USA	☑	☑
UK	☑	☑
Japan	☑	☑
France	☒	☑
Germany	☑	☑
Italy	☒	☑
CIS	☑	☒
Netherlands	☑	☑
Switzerland	☒	☑
Egypt	☑	☑
UAE	☑	☒
Saudi Arabia	☑	☒
Poland	☑	☑
Belgium	☒	☒
South Africa	☑	☑

Source: Ref. 18

entrepreneurial traders. Wholesalers / traders and supermarkets play vital roles in the distribution of organic products. As most organic production originates from small farmers, wholesalers / traders account for a 60 percent share in the distribution of organic products. Large organized producers distribute their products through supermarkets and self-owned stalls. Considering the profile of existing consumers of organic products, supermarkets and restaurants are the major marketing channels for organic products. Major markets for organic products lie in the cities viz. Bangalore, Chennai, Delhi, Hyderabad, Kolkata, Mumbai etc [7]. For the next five years, it is projected that organic spices would grow by 14 percent, fruits 8 percent and for herbal plants and cotton, it is estimated to be around 7 percent (ACNielsen ORG-MARG Survey, 2002). Markets for different range of organic agricultural products as shown in table 1.5 are estimated to reach up to 1568 tons in 2006-07.

Table 1.5. Growth forecast for specific organic products in the domestic market.

Product	Projected Growth in next 5 years (percent)
Spices (all)	14
Pepper	5
Turmeric	4.5
Tea	13
Rice	10
Fruits (all)	8
Banana	15
Mango	5
Orange	5
Pineapple	5
Herbal extracts	7
Cotton	7
Coffee	5
Oil seeds	5
Honey	5
Groundnut	5
Baby food	5
Coconut	5

Source: ACNielsen ORG-MARG, 2002

Organic farming is an eminently preferable model for the development of agriculture in India. It offers multiple benefits which include economical aspects (e.g. price premiums, high demand), natural resource conservation (e.g. improved soil fertility and water quality, prevention of soil erosion, preservation of natural and agro-biodiversity) and social benefits (e.g. generation of rural employment and corresponding lower urban migration, improved household nutrition and local food security, reduced dependence on external inputs). Furthermore, organic farming is in consonance with long-term sustainable development strategies. The spectacular growth in demand for organic agricultural products that took place in recent years has opened up export possibilities for developing countries. Aggressive marketing strategies adopted by major retail groups, the product innovations introduced by food manufacturers and supportive government policies all bode well for sustained demand growth in the sector over the medium term.

References

1. Das, S. and Biswas, B. C. 2002. Organic farming: prospects and problems. *Fert. News* **47**(12): 105-118.
2. Planning Commission, 2001. *Organic and Biodynamic Farming,* Report of the Working Group, Government of India, New Delhi, India.
3. Atkinsson, A. 1991. Parmaculture Design for Living, *In Context, No. 28,* Langley (USA), Spring.
4. Pretty J. 1998. *The Living Land- Agriculture, Food & Community Regeneration in Europe.* Earthscan Publications Ltd, London.
5. Rai, M. 2005. Organic farming: potential and strategies. 14th Millennium Guest Lecture at S. V. Agricultural College, 3rd June, 2005, Tirupati, Tamil Nadu, India, 1-13.
6. Jha, V. 2004. India. In: *Trading Opportunities for Organic Products from Developing Countries* (eds. Vossen René. and Wynen, Els). 102-123, UNCTAD, United Nations
7. Garibay, S. V. and Katke, J. 2003. Market opportunities and challenges for Indian organic products. *FiBL and ACNielsen ORG-MARG,* 1-38.
8. Lampkin, N., Foster, C., Padel, S. and Midmore, P. 1999. The policy and regulatory environment for organic farming in Europe. In: *Organic Farming in Europe: Economics Policy,* Vol. 1, Stuttgart-Hohenheim: University of Hohenheim.
9. Howard, A. 1940. *An Agricultural Testament.* Oxford University Press, p. 233.

10. Chhonkar, P. K. 2004. Organic farming and its implications on India's food security. *Fert. News* **49**(11): 15-38.
11. CAC (Codex Alimentarius Commission), 2001. Guidelines for the production, processing, labeling and marketing of organically produced foods. CAC/GL 32. 1999; Rev 1-2001. Downloaded at http://www.fao.org./organicag,frame2-e.htm. on 12 December 2001.
12. Lampkin, N. 2002. *Organic Farming*. Old Pond Publishing, United Kingdom. p. 747.
13. Willer, H. and Yussefi, M. 2000. Organic agriculture worldwide. *Stiftung Okologie & Landbau*, Bad Durkheim, Germany. p. 83.
14. Yussefi, M. 2005. Current status of organic farming world-wide. In: *The World of Organic Agriculture - Statistics and Emerging Trends 2005.* 9-18, (eds. W. Helga and M.Yussefi) International Federation of Organic Agriculture Movements (IFOAM), Bonn, Germany, 7th revised edition,
15. Kortbech-Olesen, R. 2003. Overview on world trade in organic food products: the US market and recent trends. *BioFach Congress*, Feb. 14, 2003, Nurenberg, Germany.
16. Reddy, D. N. 2001. Organic farming in India: poised for growth in the new millennium. *Proc. Fifth IFOAM Asia Scientific Conference.* Oct. 31-Nov. 4., 2001, Hangzhou, China.
17. Bhatia, U. 2001. Promotion of organic agricultural exports in India: Opportunities, constraints and policy issues. Paper presented at a workshop on *A New WTO Round on Agriculture, SPS, and the Environment: Capturing the Benefits for South Asia.* UNCTAD and the World Bank. 11-13 January, 2001, New Delhi, India.
18. Kolanu, T. R. and Kumar, S. 2004. Greening agriculture in India-an overview of opportunities & constraints. Downloaded at http://www.fao.org/DOCREP/ARTICLE/AGRIPPA/658_en00.htm#top

Chapter 2

History of Tea

The natural home of the tea plant is considered to be within the fan-shaped area included between the Naga, Manipuri and Lushai Hills along the Assam-Burma frontier in the West, through to China, probably as far as the Che-Kiang (Zhejiang) province in the East, and from this line generally South through the hills of Burma and Thailand into Vietnam. [1] A recent study, based on the geological history of Eastern Yunan and the distribution of *Camellia* species in that area indicated that the long narrow region of Wenshan and Honghe, located between $22^{0}40^{/}$-$24^{0}10^{/}$ N and $103^{0}10^{/}$-$105^{0}20^{/}$ E was the centre of origin of the tea plant. [2] However, the current distribution patterns of different varieties of tea suggest that the probable centre of origin of tea is near the source of Irrawaddy River and from there it dispersed to South-Eastern China, Indonesia and Assam.

Tea-producing countries

At present, tea is grown in more than 50 countries around the world from Georgia 43°N latitude to Nelson (South Island) in New Zealand 42°S latitude and from sea level to 2300 m above Mean Sea Level with total area under tea at around 2.77 million hectares. The world tea production in 2004 reached 3.2 million tons. Asia however, accounts for 89 percent of the area followed by Africa with 9 percent and others 2 percent. Importantly top six countries accounts for 82 percent of the production. India's share was estimated at 18 percent, based on the average from the period from 1999-2003, followed by China (24 percent), Sri Lanka (10 percent) and Kenya (9 percent). Looking at the production growth (the compound annual growth rate) across major producing countries during the period 1994-2003, India's growth rate has been only 1.30 percent compared to Vietnam's growth rate of 6.91 percent, followed by Kenya (3.47 percent), Indonesia (2.76 percent), China (2.73 percent) and Sri Lanka (2.19 percent). The area, productivity and export of tea of some major countries are given in Table. 1.1.

Spread of tea in India

Major tea-growing states in India are Assam, West Bengal, Tamil Nadu and Kerala. Tea is also grown on a small scale in a few other states viz., Tripura, Karnataka, Himachal Pradesh, Uttaranchal, Sikkim, Bihar, Manipur, Orissa, Nagaland and Arunachal Pradesh.

Although a consignment of tea seed is reported to have been brought from China to India by the East India Company as early as 1774, the history of tea in India was begun in 1823. Major Robert Bruce while visiting as member of a trading mission in 1823 discovered tea plants growing wild in some hills near Rangpur (now Sibsagar), the then capital of Assam.[3] Assam was then under Burmese occupation and in 1824 war broke out with the Burma. Robert Bruce died in the same year. His younger brother Mr. Charles Alexander Bruce was appointed in 1824 to command a division of Gunboats in upper Assam. During the progress of hostilities with Burma in 1826 he brought down certain plants and seeds which were identified as true tea *Camellia* of commerce. But there is dispute in this regard. According

Table 2.1. Area, production and export of some tea-producing countries during 2004.

Country	Area	Production	Export	Productivity world	Share of production
	(ha)	(Million kg.)		(kg/ha)	(%)
Asia					
Bangladesh	51225	56	12	1093	1.77
China	1155000	785	280	679	24.91
India	511940	820	184	1602	26.03
Indonesia	162000	165	96	1018	5.24
Japan	50200	85	9	1693	2.70
Sri Lanka	188971	309	291	1635	9.81
Taiwan	19342	21	2	1085	0.67
Turkey	76700	145	5	1890	4.60
Vietnam	84000	95	60	1131	3.02
Africa					
Kenya	132000	325	326	2462	10.32
Tanzania	21316	30	24	1407	0.79
South America					
Argentina	39000	63	66	1615	1.20

Source: Tea Board India, Kolkata; Annual Bulletin of Statistics, ITC, London.

to Baildon,[4] the tea plant of Assam was discovered by local Assamese nobleman, Dewan Maniram Dutta Barua, who later worked in the Assam Company for sometime. It might be that Dewan Maniram brought the plant to the notice of Robert Bruce during his visit to Rangpur in 1823. Further, the role of Sungpho tribe of Assam in bringing local plant to the notice of the outside world can not be ignored. It was a Sungpho Chief who supplied the seeds and plants to C. A. Bruce. This clearly shows that the Sungpho's were familiar with the tea plant.[5]

Lord William Bentinck, the Governor-General of India appointed the Tea Committee in 1834 to investigate and recommend suitable areas for tea cultivation in India. The Tea Committee reported to the Government that the tea plant was indigenous to Assam, especially in its upper reaches and about the possibility of its commercial cultivation in commercial plantings were started with local Assam tea and seed originated in China from 1836 onwards. [6] A detailed history of the growth and development of tea in India has been provided by Ukers [3] and Ghosh Hajra. [7]

An invoice of eight chests of Assam teas were sent to London in 1838 and the consignment was auctioned in London on January 10, 1839. The high prices fetched by the tea sent in London in 1838-1839 at once attracted the attentionof enterprising businessmenin Calcutta as well as in London. In 1839, the first private-owned tea company, the Assam Company was formed in London by the Board of Control and Court of Directors of the East India Company. In March 1840, practically all the East India Company's plantations, except those at Chabua, Dinijoy and two other small gardens, were handed over to the Assam Company. Later in 1852, the company emerged as the first successful tea company in the world and performance of the Assam Company attracted others to invest in tea. Later, through the pioneering efforts of local entrepreneurs, mostly Assamese, the plantation began to spread and flourish in other parts of the state. Tea plantations spread to Cachar, Tripura and Sylhet (presently in Bangladesh) from Brahmaputra valley.

Dr. A. Campbell, Superintendent of Darjeeling pioneered the experimental growth of tea in the hills in early 1841 by planting few seeds from the China stock around his bungalow which is now known as Beechwood, 2,134 meters above sea level. The first experimental trial at Darjeeling was made in 1841 with a few seeds grown in the Kumaon plantations from China stock and it was quite successful.

However, the real date of commencement of the tea industry in Darjeeling may be taken as between 1856-57. After independence, the British planters who owned about 90 percent of the estates in Darjeeling hills started selling their holdings to local entrepreneurs and by the mid - 1950, a large number of estates had changed hands. There are around seventy tea estates functioning at present in Darjeeling hills and producing 10-11 million kilos of tea every year.

A few tea estates were also opened during 1862 in Ranchi, a plateau region of Chotanagpur of Bihar state. But it could not succeed due to various reasons. After a lapse of 120 years in 1982, a small tea plantation company, i.e., Sonapur Tea Estate started tea cultivation with only 2 hectares in Sonapur – Panwara village of Pothia block in Kishanganj district. Subsequently, the cultivation spread to adjoining blocks viz. Pothia, Thakurganj and Kishanganj and it has covered nearly 1,500 hectares.

In Himachal Pradesh, the first commercial plantation was established at Holta near Palampur in 1852, at an elevation of 1,260 metres above sea level. Thereafter, the tea industry flourished and expanded its area in Kangra district and state. Britishers had started a number of tea gardens in the mid-1850s. The seeds for planting were largely obtained from China and a small quantity of China hybrid seeds and plants from Kumaon hills of Uttar Pradesh.

In Uttaranchal, tea plantation spread over 8,800 hectares, covered with the flavoury China-hybrid tea. Most of these plantations, except those in the Dehradun valley at an altitude of about 700 metres, were located far and wide between altitudes of 1,500-2,100 metres. Old tea plantations exist in the districts of Pithoragarh, Champawat, Almora, and Nainital in Kumaon, and Chamoli and Rudraoprayag in Garhwal.

One Dr. Christi experimented for growing tea in Nilgiris in 1832 and some of his tea plants were distributed for performance trial to various parts of the Nilgiri hills. In 1834 a few plants grown from the seeds brought from China, were again planted in these hills. The earliest record of commercial planting in Peermade in Kerala was in 1875. The development of Kanan Devan Hills began by James Finlay and Co. in 1878 with tea as an exclusive crop which is a landmark in the history of tea planting in this part of the country. Soon, tea cultivation caught on in Wynaad and, by 1889 planting was taken up on a large scale in the district. In the Anamallais (Coimbatore Dist.),

Box 2.1

A Quick Glance: The Tea Timeline

2737 B.C. Shen Nung, Emperor of China discovers tea. It becomes a drink of the nobility.
A.D. 729 First tea reported in Japan during Emperor Shomu's reign. Japan becomes the first country outside China to grow tea, and it becomes a vital part of the culture, finding its finest expression in the tea ceremony.
800 Lu Yu writes the Ch'a Ching, the first documented book of tea; many still regard this as the authoritative volume.
1560 Portugese make first mention of tea...thus begins a recorded contact between the Western people and tea.
1610 Dutch traders introduced tea to Europe
1664 Tea became a popular drink in England
1773 The Boston Tea Party
1780 Robert Kyd experimented with tea in Calcutta
1810 First planting of Taiwan Tea
1820 "Afternoon tea" invented by Anna, Duchess of Bedford
1823 Robert Bruce discovered tea in Bramhaputra Valley
1834 The Tea Committee was formed by Lord William Bentick
1835 Tea first planted in Darjeeling. Well this brings to light that tea is not native to Darjeeling. The high altitude and clear air was found optimum to produce a very delicate leaf, and many of the original estates are still in operation.
1838 First Assam tea arrives in England.
1839 First sale of tea in London auction (10.01.1839). Tea first planted in Ceylon
1856 Tea cultivation expanded to Nilgiris and Kerala
1869 First export of Formosa oolong
1881 The Indian Tea Association (ITA) was formed
1894 United Planters Association of South India (UPASI) was formed
1903 Tea first planted in Kenya.
1908 The tea bag invented in America
The 21st century brought innovation to tea that included Instant tea, canned tea and flavored tea, much against the skepticism of tea purists.

the actual opening of tea estates was around 1897. Karnataka came on to the tea map, rather recently.

Today, India has 39,700 tea estates (32,000 in the south and 6,700 in the north) and a tea-producing workforce of more than a million people. India is the largest producer and consumer of tea. Almost all of the teas produced are 'black tea', although a small quantity of green tea is also produced. The Darjeeling tea is famous for its muscatel flavour. Assam tea is granular, grayish to black in colour. This tea has strong bright liquor.

Tea production scene in India

Post-independence era witnessed Indian tea production making decent progress. From around 278 million kg during 1950s, Indian tea production grew by leaps and bounds to touch over 850 million kg by the beginning of the 21st century. Of the total tea produced in the country during 2004, North India contributed the most with a share of 77.4 percent; the remaining 22.6 percent being accounted for South India.

Other major tea-growing countries

China

The history of tea cultivation in China can be traced back to three thousand years ago. The tea plantations in China are located in between 94 ^{0}E and 122 ^{0}E longitude, and from 18^{0}N to 37^{0}N latitude, on elevations up and 2600 m. But the areas for economic cultivation are concentrated in the region between 102 ^{0}E Eastwards and 32 ^{0}N Southwards. The climate of tea areas belongs to the humid monsoon climate of the East coast type. In the subtropics of China, the climate is not as dry as the general subtropics: it is moist, has a lot of precipitation and high temperature in the summer.[8] The tea-producing areas are distributed in 19 provinces viz. Anhui, Jiangxi, Zhejiang, Jiangsu, Hunan, Sichuan, Guangxi, Yunnan, Guizhou, Guangdong, Hainan, Hubei, Taiwan, Fujian, Shanxi, Shandong, Gansu, Henan and Tibet. The four major tea-producing provinces are Zhejiang, Hunan, Sichuan and Anhui together account for 60 percent of the national output. The tea areas are divided into 4 parts, namely the South China tea area, the Southwest China tea area and the South of Yangtse River tea area and North of Yangtse River tea area. China for a long time held a pre-eminent position in tea production and export. The target proposed various kinds of tea being produced in China. According to processing methods and technologies, Chinese teas are basically divided into six categories. These are green tea, black tea, oolong tea, white tea, yellow tea and black-black tea. Of these, green tea forms the main bulk (more than 60 percent) and black tea accounts for about 20 percent of total production. The main green tea varieties are fried green tea, with 10 sub-varieties: pan-fried dried and sun-dried; steamed green tea, with 2 sub-varieties and "origin green teas" with 14 authentic origins. Oolong tea is a semi-fermented tea and mainly made in Fugian, Guangdong and Taiwan provinces. White tea, a light fermented tea primarily made in Fujian province. Black-black tea is a post-fermented tea and consumed mainly by the people living in remote border areas of China.

Sri Lanka

Dr. N. Wallich of Calcutta Botanic Gardens sent Assam tea seeds to Sri Lanka (the then Ceylon) and seeds were received at the Royal Botanic Gardens at Peradeniya near Kandy in 1839. Thereafter, the Government of Sri Lanka received 200 plants in 1841 which were planted in Nuwara Eliya. Two further consignments of plants were received in early 1841 and 1842 from China and Assam, and were planted in Rothschild estate at Udapusselawa and Pennyland estate, Dolasbage respectively. Planting of tea in Sri Lanka in a serious way began in the decade of the 1880s. Tea in Sri Lanka was essentially a replacement of coffee which was fast dying out because of attack of the dreaded coffee rust (*Hemileia vastatrix*). Tea in Sri Lanka is produced under varied ecological regimes and for easy reference these are grouped on the basis of elevation into three distinctive broad categories of low (below 600 m), medium (between 601 and 1200 m) and high elevation (above 1200 m) tea areas. The entire production of tea is in the form of black tea and almost 98 percent is manufactured by orthodox method. Plucking continues throughout the year. Currently, Sri Lanka is the leading exporter of tea in the world.

Japan

Japan is among the major producers and consumers of tea, the fourth-highest volume in the world after India, Russia and Britain. There are about 51,000 hectares of tea fields in the country that yielded 85,000 tons. Known for its consumption and export of green tea, Japan imports a wide variety of tea from around the world. The Japanese call the green tea o-cha and has numerous varieties that differ in the cultivation and the processing methods. Sencha, Gyokuro, Macha, Kamairicha, Bancha, Hojicha are some varieties of green tea that Japan grows. Not only used as a beverage, the Japanese also put green tea as an ingredient in ready-to-drink teas, soft drinks, ice cream, biscuits and lollies. Green tea has been recognised for its digestive capabilities as well as for its flavour and soothing aroma. Sencha is fresh, grassy, refreshing and cooling. Sencha tea is bright green or lighter, with the higher quality teas infusing greener and tasting more vegetal than the lesser quality teas, which are paler and grassier. Sencha Honyama, is pale green in colour and has a clean, flowery flavour that makes it especially relaxing and soothing. Gyokuro green tea is the best green tea Japan produces. Known as pearl dew, Gyokuro has high chlorophyll content, low tannin and a leaf with dark emerald green colour. It is also so finely processed that makes it powdery. Matcha Uji, is a special tea made for the famous Japanese tea ceremony called Cha-no-yu. Made from the Gyokuro tea plant, the brewing is done in the traditional way. Bancha Houjicha is crisp and nutty-tasting because the leaves are roasted. This tea is light brown and is served to accompany noodles and sushi. Genmaicha is made by mixing green tea with roasted rice and popcorn. This innovation was made during the Second World War to meet the tea shortage. Today Genmaicha is a favourite Japanese treat. The origin of Japanese tea can be attributed to China. Monks who studied in China in the 11-12th centuries brought tea to Japan. Today, drinking tea is a ceremony in Japan, which has become a part of the Japanese tradition.

Indonesia

Tea in Indonesia was found by the Dutch in the 18th century. In those days, the island of Java was used as a staging port by the Dutch East India Company for the onward transportation of Chinese tea. By the early 18th century, Dutch settlers were already establishing tea plantations on the island. Production soon followed in Sumatra and later on the island of Sulawesi. The Second World War saw a decline in the tea business that was a way of life for the Indonesians for the past 200 years. In 1984, Indonesia's tea industry was revived with much effort and investment. Since the establishment of the Indonesian tea board, constant improvement and implementation of modern methods of tea production have taken place. Today Indonesia produces 165,000 tons of tea. Indonesian tea differs from other tea-producing countries in respect to location, soil, and the climate where the tea estates are found. Teas there are planted in the highlands where volcanic soil and tropical climate are predominant. The main product is the black tea and about 80 percent of production is exported. Green tea is also cultivated in Indonesian plantations. The bulk of Indonesian tea is grown on the island of Java. The best quality comes during the dry season of August and September, though tea is picked the whole year. The best quality tea comes from government-owned estates. Black tea production is currently based on tea plants originally imported from Assam, and are mostly exported for blending. Indonesian teas are light and flavourful. Indonesia is the only country where tea auction is not conducted in English.

Taiwan

According to the history of Taiwan by Mr. Lien Ya - Tang, one Mr. Ko Chao brought a few tea plants to Taiwan from Wu-i, Fukien Province, China and planted them in North Taiwan during the Chia - Cheng period (1796 - 1820). In fact, it was the beginning of tea cultivation in Taiwan. However, commercial tea cultivation began in Northern Taiwan using seeds from tea plants brought by Chinese mainlanders from Fukien Province about two hundred years back. In 1865, the first export of 82 metric tons of Taiwan oolong tea was officially recorded. Later, John Dodd, an English businessman was exporting 121 metric tons of oolong tea to the United States. In 1881, Wu Fu - Lao, a tea trader from Fukien came to Taiwan to produce Paochung tea and Taiwan started to export Paochug tea [9]. Thereafter, in 1885, Liu Ming - Chuan the first Chinese Governor of Taiwan appointed by the Ching Dynasty began to develop the tea industry systematically and the total tea export in 1893 increased to more than 9,800 metric tons.

Tea areas range from flat land to high hills with an altitude of 2,600 metres. The tea plantations were found in Taipei, Taoyuan, Hsinchu, Miaoli, Nantou, Chiayi, Yunlin, Yilan, Hwalien and Taifung counties. But 50 percent of tea was produced in Nantou, Taipei, Chiayi and Taoyuan areas.

Malawi

In 1878 Jonathan Duncan, a gardener at the Church of Scotland Mission in Blantyre, obtained seeds from the botanic garden in Edinburgh but these seeds failed. Another batch of seeds was brought by Dr. Elmslie in 1886 from Edinburgh, as well as a few from Kew Gardens. Mr Duncan tried again and obtained just two bushes which were planted in the grounds of the Mission in Blantyre. One of these bushes became the ancestor of the original tea plantations in Mulanje and vegetatively propagated bushes from this original bush are still growing in the Tea Research Foundation. During 1891, Mr. Henry Brown procured a few seeds from Mr. Duncan and half of it were planted in his estate (Thornwood) and the other half on a neighbouring estate (Lauderdale) owned by John Moir. Thereafter, tea estates from 1908 onwards were opened in the Mulanje (700 m to 1000 m elevation) and Cholo (900 m to 1300 m elevation) districts in the south. Initially planting was done with China hybrids but later

Assam jats were introduced. Expansion of tea cultivation then spread over to the northern district of Nkhata Bay in 1960. With its 45,000 mt production, tea is Malawi's second largest export, after tobacco. The biggest export destinations are the UK and South Africa. Their superb colour and brightness lends them admirable use in the blending of leading British tea brands. Today, large commercial estates account for 93 percent of production, with the remainder grown by some 6,500 smallholders.

Kenya

Tea was introduced in Kenya in 1903. Production of tea in Kenya was virtually confined to estates until the 1960s. Then came the special Crops Development Authority (set up in 1960) to develop smallholders' tea production. The Kenyan Government established the Kenyan Tea Development Authority (KTDA) in 1964 to develop tea among the small holder and also to extend cultivation to all suitable areas. About 400,000 smallholders, grouped under the KTDA, produce 62 percent of national production, and the rest by commercial estates such as those operated by Brooke Bond, Unilever and James Finlay. The principal tea-growing districts of Kenya are located in an area lying between Mount Kenya and Lake Victoria on either side of the Rift Valley. The tea estates are located on either side of the equator. The main tea-growing district is Kericho in the west followed by Nandi, Sotik, and Kissii around Kericho. Planting is done at an elevational range of 1,600 m to 2,250 m. Kenya produces only black tea. About 90 percent is of CTC manufacture. Between 8 - 15 percent is used for local consumption and the rest is exported. The Kenyan tea is very bright, colourful, with a reddish coppery tint and a pleasant brisk flavour. Kenya's production in 2004 was 310,000 mt and the main buyers are Pakistan, UK and Egypt.

Tanzania

Tea was first planted in Tanzania in 1902, when German settlers introduced the crop in the Usambara Mountains at the Agricultural Research Station at Amani and at the Kyimbila Mission in the Rungwe district near Tukuyu in the Southern highlands. Commercial production began in 1926 in Usambaras and Njombe. In 1934, Tanzania produced 23 tons of made tea. After World War II, when the British took over the tea plantations, production rose from 2,723 hectares in 1945 to 7,336 hectares in 1960, yielding 3,722 tons of made tea.

Before Tanzania's independence in 1961 large estates dominated tea production. By the mid-1960s the government was encouraging smallholder production, with some success, and by 1985 smallholders accounted for almost 30 percent of output. In 1968, Tanzania Tea Authority (TTA) was established by an Act of Parliament as a state-owned corporate body. TTA has its headquarters in Dares Salaam. It manages a number of estates and operates tea processing factories despite supervision of marketing of tea. Most of the smallholder green leaf went to the eight Tea Authority-owned factories for manufacturing, and the rest to factories owned by the estates. It may be noted that smallholder tea production in Tanzania has been supported by two World Bank operations: a $1.4 million loan in 1966 (part of an agricultural credit project) and a $7.1 million loan in 1972. By the late 1980s, however, there were visible signs of distress in the smallholder sector. By the mid-1990s its share dropped below 10 percent and by the 1998 it fell to 5 percent. Estate production, on the other hand, followed a large independent path, with output growing considerably during 1990s. By the 1995/1996 season, estate yields were 10 times those of smallholder plots.

Tea contributes about $25 million to Tanzania's export earnings, making it the fifth largest export crop after cashew, coffee, cotton, and tobacco. More than three-quarters of Tanzania's tea are exported. Currently, exports go primarily to the UK and Pakistan. Tea provides employment to 50,000 families and directly or indirectly affects as many as 2 million Tanzanians. Though tea is economically important for Tanzania, the country produces less than 1 percent (25,000 tons) of world tea production.

Consumption of tea

India and China are the two largest tea consuming countries, accounting for 20 percent and 16 percent, respectively, while the United Kingdom, the Russian Federation, Pakistan and Japan each account for about 5 percent of the global total. Tea is a staple beverage for the Indian customer with a yearly per capita consumption of around 660 grammes (Table 2.2). Current per capita consumption has been growing at about 1.6 percent annually in line with the population growth rate. The domestic consumption figures have been grown from 415 million kg in 1986 to 640 million kg in 2004. In terms of growth rate (CAGR) during the 15-year period, tea consumption grew at 2.89 percent, with packet growing at an impressive 5.6 percent and loose tea at 1.63 percent.

Table 2.2. Apparent consumption of tea in major producing and importing countries.

	Consumption per head, kg
Producing Countries	
India	0.66
Sri Lanka	1.38
Japan	1.14
Kenya	0.32
Tanzania	0.08
Turkey	1.98
Iran	1.30
South Africa	0.45
Importing Countries	
Europe	
U.K.	2.24
Germany	0.28
France	0.23
The Netherlands	0.44
Irish Republic	2.96
USSR/CIS	0.79
Sweden	0.30
Italy	0.09
Poland	0.82
America	
U.S.A.	0.33
Canada	0.61
Chile	0.99
Asia	
Iraq	2.42
Hong Kong	1.37
Kuwait	2.29
Qatar	2.89
Pakistan	0.75
Afghanistan	1.84
Syria	1.62
Africa	
Morocco	1.42
Tunisia	1.13
Egypt	0.90
Libya	2.92
Sudan	0.44
Oceania	
Australia	0.70
New Zealand	0.95

Source: Annual Bulletin of Statistics, ITC, London; Consumption per head is calculated on the total population as at the middle of each period, does not take into account differing composition by age and race. No allowance is made for consumption by visitors from other countries.

Exports

Asia (other than West region) accounts for a major share of 57.17 percent of world demand. It is followed by West Asia (13.2 percent), East Europe (8.08 percent), European Union (6.69 percent), North Africa (5.12 percent), America (4.48 percent), Africa other than North (4.14 percent) and Oceania (0.73 percent). On analyzing the region-wise demand for tea for the period from 1997 to 2000, the growth (CAGR) in demand was as presented in Table 2.3.

Table 2.3. Growth (CAGR) in demand (%) for tea in importing countries (1997 - 2000).

Region	CAGR
EU	-3.53
Other Western Europe	1.09
East Europe	-0.26
America	1.42
West Asia	3.48
Asia other than West	2.65
North Africa	-2.17
Africa other than North	-0.63
Oceania	-2.45
World	1.55

Most of the North Indian orthodox teas cater to the demand of the American and European markets, South Indian CTC teas goes to Russian Federation, Khazakistan and West Asian markets. Of the 180 million kg of tea exported during 2001 (Table 2.4), 102 million kg was of CTC while the rest 78 million kg was orthodox. During the year '96 million kg of orthodox teas was produced in India, of which 81 percent found their way to overseas markets.

India, which commanded the first position in the export market a few years back, is now pushed to the fourth place with an estimated share of 15.6 percent in 2000 of the total export market, way behind Sri Lanka (20.8 percent), Kenya (18.57 percent) and China (18.50 percent).

While the global export market for tea has steadily grown, share of India has declined from over 50 percent in 1950s to mere 15 percent by the end of the 20th century. It further declined to 12.24 percent in 2004, much to the advantage of the new entrants' viz. Sri Lanka, Kenya, Indonesia and Vietnam. The stagnating export figure at around 200 million kg since the last five decades is a clear reflection of India's dismal performance in tapping the growing global demand.

Table 2.4. Share of India in global tea export market.

Year	Indian exports (Million kg)	World exports (Million kg)	Percent share of India to world export
1950	200.78	398.98	50.32
1960	193.06	530.00	36.43
1970	202.33	651.10	31.08
1980	224.78	858.90	26.17
1990	210.02	1134.96	18.50
2000	206.81	1325.32	15.60
2001	182.59	1398.00	13.06
2002	201.00	1434.00	14.02
2003	173.68	1300.00	13.36
2004	183.64	1500.86	12.24

Source: Tea Board of India; Sugar and Beverages Group, FAO, Rome; ITC Annual Bulletin, London.

Health value of tea

Tea is a refreshing, thirst-quenching, inexpensive, versatile, safe, virtually calorie and sodium-free beverage when had without milk and sugar. Tea contains carbohydrates, amino acids, proteins, lipids, vitamins and minerals.

Tea has always come with a positive health message ever since the Chinese started drinking it. The monk Eisai, considered the "Father of Tea" in Japan, wrote in his book 'Maintaining Health by Drinking Tea' in 1211: "Tea is a miraculous medicine for the maintenance of health. Tea has an extraordinary power to prolong life". But in those days the health claims could not be substantiated by scientific evidence. Recent studies in leading medical journals declare tea a potential heart tonic, cancer blocker, fat buster, immune stimulant, arthritis soother, virus fighter and cholesterol detoxifier [10, 11]. However, more comprehensive, in-depth cohort studies are needed to thoroughly evaluate the association between tea drinking and risk of chronic diseases.

Tea drinking around the world

Tea is served and drunk in a number of different ways in India. Sometimes it is brewed and served with milk and sugar, or leaves are boiled with milk, water, spices, and sugar. On railway stations and trains, streets, corners, sweet milky tea is poured from hot kettles into disposable cups or mugs.

Tea is always served to guests in modern Chinese homes as a sign of welcome and hospitality. They serve a range of different tea which is brewed in glasses, covered mugs or little Chinese tea pots and bowls. The love of tea spread to Japan from China with Buddhist monks who traveled to and from China. The modern Japanese society has kept alive the famous 'Cha-no-yu' or the tea ceremony from the early days of the Zen Buddhist monks. The tea served in the tea ceremony is called *matcha* (pronounced 'mahcha') and is powdered tea of the highest quality. Matcha is beaten with lukewarm water with a whisk in a small porcelain bowl and served. The Japanese believe that "tea is more than an idealization of the form of drinking and it is a religion of the art of life". Today, Japanese mostly drink green tea but there is a noticeable trend to turn to black tea among certain sections of the Japanese.

In Tibet, green tea is broken up and boiled in water then strained into a churn and mixed with goats or yak's milk, salt, and butter.

In Iran and Afghanistan, people drink green tea as a thirst quencher and black tea as satisfying warming refreshment from little porcelain bowls. In Egypt, tea without milk is brewed strong and served in glasses, and sometimes flavoured with sugaror mint leaves. And in Morocco, black tea is brewed with sugar and mint in a long-spouted pot and is then poured in a thin stream into tall glasses.

In Turkey, black tea is served in curved glasses and offered on trays to visitors throughout the day. Russians drink both black and green teas usually from glasses. They generally take sugar cube or a spoonful of jam before sipping the tea.

Like British and Indian, most of the Africans drink strong tea with milk. Tea is also a widely consumed beverage in America. Iced teas comprise about 80 percent of all teas consumed in the United States. But the fashion of drinking hot black tea with milk is now growing. The 'Gourmet' or 'Speciality Teas' is getting popular among the elite Americans. A wide variety of speciality shops (nearly 1,500 outlets) and mail order companies are offering a carefully chosen range of teas from all over the world. Much of this demand is triggered by news of the health benefits of tea and the desire of consumers to make tea a regular part of their diet.

References

1. Harler, C. R. 1966. *Tea growing*. Oxford University Press, London. p. 162.
2. Yu, F. 1986. Discussion on the originating place and the originating centre of tea plant (in Chinese). *J. Tea Sci.* **6**:1-8.
3. Ukers, W. H. 1935. All about tea. *Tea and Coffee Trade Journa*l *Co.*, New York, USA
4. Baildon, S. 1877. *Tea in Assam*. Calcutta.
5. Barua, D. N. 1989. *Science and practice in tea culture*. Tea Research Association, Calcutta - Jorhat, p.493.
6. Griffith, P. 1967. *The History of Indian Tea*. Weidenfield and Nicolson, London. p.730.
7. Ghosh Hajra, N. 2001. *Tea Cultivation: Comprehensive Treatise*. Int. Book Distribution Co., Lucknow, p.518.
8. Huang, S. Meteorology of the tea plants in China: A review. *Agric. Forest Meteorol*, **47**:19-30.
9. Chiu, Tsai-Fua. 1991. The present status on tea cultivation, manufacture and utilization in Taiwan. *Proc. Intl. Symp. On Tea Science*. Shizuoka, August 26-29, Japan. 34-39
10. Hara, Y. 2001. *Green Tea: Health Benefits and Application*. Marcel Dekker, Inc., New York, p.252.
11. Ghosh Hajra, N. 2006. *Tea and Health: Science Behind the Myths*. UBS Publishers' Distributors Ltd., New Delhi, p. 178.

Chapter 3

Production of Organic Tea in Different Countries

Tea qualifies as organic only when there is active use of environment-friendly techniques and it is certified to be so by the accredited inspecting authority. The number of organic tea producers and the volume of organic tea traded on the world market have increased substantially over the last few years. This development can be explained by a number of factors. In the first place, tea farmers have become more aware of environmental problems (erosion, pesticide residues in tea plants) and severe health hazards connected with an intensive system of tea production. A further reason for the rise in organic tea can be explained by the fact that the demand for organic tea has grown constantly as a result of increased consumer awareness of pesticide residues and heavy metals in conventional teas. Furthermore, there is belief that organically grown teas are generally of better quality due to the avoidance of artificial additives.

Area and production

India and Sri Lanka are among pioneers for growing organic tea. It has been grown in these countries for more than 15 years. In India, the Darjeeling district of West Bengal has seen many gardens undergoing conversion to organic. Prime movers in this district are Makaibari, Ambootia, and Chamong. Currently, the production of organic tea increased more than 23 times compared with that of 1990. A list of tea estates in India that has taken up organic cultivation is presented in Table 3.1.

Organic tea production in India started in Makaibari tea estate located in Kurseong in the district of Darjeeling, West Bengal state. Since 1986 it has been cultivating organic tea according to the guidelines of the International Federation of Organic Agriculture Movements (IFOAM) and Demeter International, Germany. Out of total plantation area of 670 ha, 430 ha are under primary subtropical rainforest. In 1994, Ambootia tea estate (Fig. 3.1) was brought under organic cultivation according to the guidelines of IFOAM. The tea of Ambootia was certified in 1997 by the Institute for Marketology, Switzerland; NATURLAND - Verband, Germany and Bio Suisse, Switzerland. Selimbong is another organic tea estate, located at an altitude of 1,400 m. With 160 ha area under tea, Selimbong produces exquisite Darjeeling teas. The garden has been certified organic by Naturland, Germany since 1995. These estates adopted Biodynamic practices and are certified Biodynamic by Demeter International, Germany since 1998. Poabs estate in Nelliampathy hills, Kerala is also growing tea following biodynamic practices. Currently, there are about seven tea estates in India that adhere to the prescribed on-farm process and are "certified biodynamic', although some of the biodynamic preparations are used in varying degrees on organic tea estates across India.

Table 3.1. Organic tea estates/projects (certified and under conversion) in India.

Darjeeling

Ambootia, Arya, Chamong, Edenvale, Mohon Mazua div. of Goomtee, Mahalderam div. of Jungpana Makaibari, Monteveiot, Mullootar, Nagrifarm, Pussimbing, Samabeong, Seeyok, Selim Hill, Selimbong, Singell, Singtom/Steinthal, *Tumsong, Samjukta Vikash Cooperative (small growers project),* Avangrove, Sungma, *Turzum, Mim, Dhajea*

Dooars

Pathurjhora, Kamala and sections of Dalgaon tea estate

Assam

Bherjan, Chowkhani, *Fatemabad, Hokonguri, Koombergram,* Maud, *Mokalbari,* Sewpur, *Sonarie, Banaspaty,* Ramanugger, *Ramsa Hill, Rani,* Rembeng-Kondoli, *Kopili,* Jutlibari, Baghmara and Na-gaon divisions, Madoorie, *Sankar,* West Jalinga, Monkhooshi

Nagaland

Area of Zunebhoto, Wakshing, Psuengfang, Dimapur, Mokokchung, Melourie and Phek near Khohima and Wokha district

Tripura

Fatikcherra

Arunachal Pradesh

Abali, Jia, Koronu

South India

Oothu, Korakundah, Dunsandle (partly), Chamraj, Chundavurrai and Chittavurrai, Devabetta, Iyerpadi, Manar (partly), Palmyrah, Toramala *(partly), Thuthampara, Seetharagundu, Stagbrook,* Welbeck *(Partly),* Poabs, Nallathany, PDS-small growers, Thiasola, AV George tea estate

Uttaranchal

Arcadia, Hurbanswala, East Hope Town

Tea estates in italic are certified by IMO; underlined by SKAL; Avangrove and Sungma tea estates in Darjeeling by SGS; others not known, Total area under organic tea (Certified) in India is 4,500 ha.

Around 500 ha tea cultivated in virgin forest land in Nagaland, mostly by small growers, where no chemical inputs were applied prior and after cultivation. Since there is no tea processing factory in Nagaland at present all teas harvested are hand made tea and marketed domestically. International organic certification process for the produce is in progress.

In Arunachal Pradesh, 33, 10 and 12 hectares in Abali (partly), Jia and Koronu tea estates respectively planted in 1996 to1997 and converted to organic during 2004. Indocert is the certifying agency of these estates. Despite these tea estates, there are number of small growers in Arunachal Pradesh who do not use chemicals and are organic by default

Fig. 3.1 Ambootia biodynamic tea estate in India.

In Darjeeling, first of all, one of the major reasons for tea gardens to turn organic was because yields were decreasing. For example, the total Darjeeling black tea production was about 14 million kg in 1980 on an area of 20,000 ha, but currently it is lesser than 10 million kg (Table. 3.2). The decrease in yield was not merely due to excessive usage of chemicals and fertilisers, but also because of the fact that most of the tea bushes are over 100 years old and need replanting. Secondly, in the beginning of the 1990s restrictions from the export markets on chemical residues implied that most of the tea gardens had to substantially reduce the usage of pesticides and chemicals. Hence, moving from reduced use of pesticides and

Table 3.2. Region-wise production, area, productivity of conventional and organic tea in India in 2003.

Region	A	B	C	D	E	F	G	H	I
Assam	453936	270163	55.3	1691	2997	2276	1316	36.2	22
Darjeeling	9,742	18000	1.2	541	1900	4300	442	22.9	18
Dooars	130739	70200	16.0	1862	249	457	545	3.0	129
South India	203116	115080	24.8	1765	3000	2800	1071	36.3	39
Tripura	6508	6700	0.8	971	42	120	350	0.5	64
Uttaranchal	2694	8800	0.04	306	81	600	135	1.0	56

A= Conventional tea production (mt); B= Total area (ha); C= Percent over Indian production; D= Productivity (kg ha^{-1}); E= Organic tea production (mt); F= Total area under organic tea (ha); G= Productivity of organic tea (kg ha^{-1}); H= Percent over Indian organic tea production; I= Percent loss of organic yield over conventional yield.

Organic tea includes production of certified and under conversion area. India's certified organic tea production reached 3,000 and 3,500 mt in 2002 and 2003 respectively. Total tea production 82,0000 mt and organic tea production in India 8269 mt (Source: www.indiateaportal.com; Field survey, and secondary sources: Agricultural and Processed Food Products Export Development Authority (APEDA), Tea Board of India and by collating information from different internet sites) chemicals to organic tea was not seen as a big step forward. Thirdly, in the beginning of the 1990's market premium of over 80 percent were realised for organic tea. This prompted many growers to begin exporting organic tea. Most importantly, application of chemical inputs in the tea estates perhaps led to soil and other environmental problems in Darjeeling hills. This led to considerable destruction of property and some converted to organic agriculture as a response to these environmental problems.

The organic teas are packaged under stringent international quality control standards - using ecofriendly material like brown kraft paper, hand-made paper, recycled board and natural jute fibre. Tea bagging use unbleached filter paper. Makaibari, Ambootia, Selimbong, Samabeong etc gardens are registered with Fair Trade Labelling Organisation International (see Box 3.1) and meet their various standards of worker welfare. In Makaibari tea estate, several social and economic development projects with the participation of their workers for their economic prosperity and increased motivation for the development of the garden were initiated. The programme includes empowerment of women (promoting women in supervisory positions), micro-loan programmes, improvement of sanitation facilities, education, family planning, free childcare, participatory afforestation, and biomass production etc. Many displaced workers in Darjeeling rehabilitated through Samjukta Vikash Cooperative (small growers' project) by growing organic tea. Tea Promoters India is associated with various small farmers initiatives. The workers of the abandoned Darjeeling tea estate 'Mineral Springs' are getting technical help from the Tea Promoters India. They are providing technical know-how, and arrange certification for the small growers. Green leaves of the small farmers of the Mineral Spring are processed in Selimbong tea estate factory. The Fair Trade organisation, Gepa, has been marketing the whole harvest of the Samabeong tea garden for a long time. Gepa has a plan to launch new tea from the small holder project "Subarna" in 2005, celebrating its 30 years anniversary. Tea Promoters India trained the small farmers living around the Samabeong tea garden in organic tea production. The plucked tea leaves will be processed in the tea factory of Samabeong. The small farmers have now the chance to get additional income for their families besides their vegetable production (http:// www.gepa3.de).

In Assam's Karbi Anglong district, close to the border of Nagaland state, Banaspaty tea estate is located in the tribal heartland.

Madoorie tea garden is isolated from conventional garden on all four sides and is divided into two sections by a river. It is situated in the high quality districts of Jorhat. Approximately 175 hectares are under tea cultivation, which produces orthodox/CTC varieties. Rani is another organic tea estate under MKB (Asia) Pvt. Ltd., located in a hilly area of Assam. Approximately 300 acres of the 900 acres are under tea cultivation while over 500 acres of land is covered with lush rainforest. The Institute of Marketology (IMO) based out of

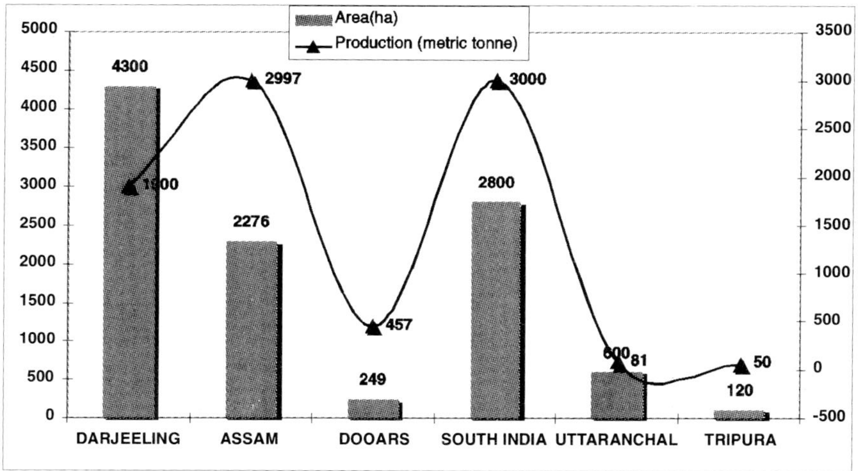

Fig. 3.2. Area and production of organic tea (certified by intenational agencies and under conversion) in different regions of India.

Switzerland is the organic certifying agency for Rani. This estate started conversion to organic in March 2002. The plantation converted to organic using biodynamic agricultural methods. Rani hosts an extensive network of fisheries with 23 acres under aqua cultivation. The plantation also hosts a diary farm and a piggery. Rani tea estate is self-sufficient in meeting its energy needs. A mini-hydel power plant was constructed that was co-sponsored by United Nations Development Program which powers all of Rani's operation. The plantation produces both orthodox and CTC black teas, and also green and white teas. The factory produces around 300 mt of made-tea each year.

In South India, the United Nilgiris Tea Estates Co. and Bombay Burmah Trading Corporation(BBTC) were among the first companies to convert some of their tea and coffee estates to organic; more recently, some areas in Tata Tea, a dominant producer in Munnar region of Kerala, has already begun growing organic tea in its 250 hectares plantations mainly for supply to the foreign bulk buyers in Europe and the US. The company will launch tea under the International brand Tetley. In 1988, BBTC in Tamil Nadu decided to put 39 hectares in Manjolai area under organic cultivation on a trial basis. An importer in the Netherlands lent assistance by contracting certifying agencies for the project. By 1989, a decision had been taken to put the whole of Oothu tea estate under organic cultivation, and full-scale conversion, which took three years. Oothu tea estate (under Singampatti Group of Estates of the BBTC) was planted with several clones and certified for organic in 1991 -1992 by IMO. The estate was brought under biodynamic (BD) cultivation practices in 1997 and has been certified by Demeter. Today, BBTC has 312 hectares of tea under organic cultivation, an area with the capacity to produce 900 tons certified organic tea. BBTC is the largest single producer of certified organic and biodynamic tea in the world. Bombay Burmah has a dedicated factory for production of a range of organic and bio-dynamic teas which include black orthodox, CTC teas and green teas, depending on the buyer's requirement. Korakundah is another tea estate situated at an elevation of around 1,725 meters in Tamil Nadu (Nilgiri) produces organic green and black teas certified by IMO, and are also 'Kosher" certified. Dunsandale garden situated in the heart of Ootacamund in the Nilgiri Hills comprises 155 hectares and produces orthodox teas. The Poabs organic estate in Kerala, at an average elevation of 1,200 meters, growing coffee and tea, is now identified as the single largest perennial multi-crop organic plantation in the world. First opened in 1889, the Seethargundu estate went through several ups and downs over the years and it was even lying abandoned for 16 years before the Poabs Group took over the estate in 1989. Conversion to biodynamic agriculture commenced in 2000, under the supervision of New Delhi-based biodynamic consultants Natura Agrotechnologies. Today, 350 cows are maintained on the estate, along with a nursery growing the herbs required for BD preparations and extensive compost yards.

Peermade is a small town in the mountains of Kerala in Southern India. Over 1,000 small growers under Peermade Development Society (PDS) are producing organic tea. A factory, Sahyadri Small Farmer Tea Factory, with a production capacity of 600 tons organic made tea per year was inaugurated in 2003. The organic tea factory is the first joint project of Peermade Development Society and Naturland Association in Germany. The training for the small growers was organised by PDS and supported by Naturland. The growers founded the Sahyadri Farmers Consortium and are herewith the owner of the organic factory.

In Tripura, Fatikcherra tea estate is of 120 hectares and produces 42 mt. of CTC/Othodox/green teas. This estate certified by SKAL International.

In Uttaranchal, M/S Dehra Dun Tea Co. Ltd., comprising Arcadia and Hurbanswala tea estates were established in the year 1863. Both are producing organic green teas duly certified by SKAL International based in Netherlands. An area of 600 ha is under organic tea and total production of organic green tea at present is around 81 mt. in Uttaranchal (Fig. 2.2).

In Nagaland, mostly by small growers, around 500 ha tea cultivated in virgin forest land where no chemical inputs were applied prior and after cultivation. Since there is no tea processing factory in Nagaland at present, all teas harvested are hand-made tea and marketed domestically. Construction of three factories, one each in Zunebhoto, Dimapur and Mokokchung are in progress. The organic tea plantations in India are collectively produce under the umbrella of the 'Indian Bio Organic Tea Association' (http://www.snonline.com/ibota). The logo of Indian organic farmers' association is presented in Fig. 3.3.

Fig. 3.3. Level of Indian Organic Farmers' Association.

Organic tea is the single largest organic product in Sri Lanka and has a major share among the exported agricultural commodities. It accounts for 26 percent of the organic food export turn over, which was valued at US$ 3.5 million in 2001 (Source: President of Lanka Agricultural Movement, Hill Crest Green, Bowalawatta, Udaperadeniya, Sri Lanka)

Currently, the production of organic tea in Sri Lanka is around 1,200 tons (Source: personal communication with Dr. M. Rajsingham, Sri Lanka, 26 April, 2005). Stassen Natural Foods Company, Sri Lanka which is a part of the Stassen Group, commenced with cultivation of organic tea in 1985 and started with seven gardens certified by Naturland of Germany (Table 3.3). Today, 10 gardens have converted to organic production in Sri Lanka. The tea country is divided in Sri Lanka into a Western and an Eastern region by a chain of mountains which forms the climatic divide of the island. On the East is the Uva

Table 3.3. Area and production of important organic tea gardens/projects of Sri Lanka in 2002.

Estate/Assoc.	Organization/Company	Tea area (ha)	Production (tons)
Iddalgashinna	Stassen Natural Food Ltd.,	353	206
Ventura	Stassen Natural Food Ltd.,	290	300
Needwood	Needwood Emmag	94	70
Greenfield	Lanka Organics Pvt. Ltd.,	70	58
Koslanda	Maskeliya Plantations Ltd.,	160	80
Smallholder Co-operative	Gami Seva Sevana Ltd.,	83	7
Small Organic Farmers' Assoc.	Bio Foods Ltd.,	214	67

Source: Swiss Import Promotion Programme (SIPPO)

region and on the West the Dimbulla region, each producing teas renowned for their distinct character and flavour. Ventura is Sri Lanka's one of their biodynamic gardens situated in the Bogowanthalawa Valley in the Dimbulla region and its elevation ranges from 1,100 to 1,300 metres. This estate is famous for the production of premium quality organic black teas, rich in flavour and colour. Iddalgashinna is another Stassen's biodynamic tea garden (Fig. 3.4) in Haldemulla in the Uva district producing organic black and green teas. Its elevation is from 1,000 to 1,900 metres. In 1987, Iddalgashinna garden achieved distinction as the world's first certified organic tea garden. Needwood tea garden is certified by IMO, Switzerland and the National Association for sustainable Agriculture Australia (NASSA). Needwood was the first organic tea plantation in the world. The Original Ceylon Tea Company is offering this fine Needwood tea to North America. Lanka Organics' tea is also grown in the Greenfield tea estate located in the Uva highlands. This is a certified level 'A' (full organic) plantation and is included on the International Fair-trade Tea Register. Greenfield tea

Fig. 3.4. Iddalgashinna garden in Sri Lanka

estate is certified by IMO, NASAA (Australia), SKAL (The Netherlands). Organic Farmers & Growers (UK) also inspecting annually. Koslande is another organic tea estate under Maskeliya Plantations Ltd. is located in the Uva district.

M/S Gami Seva Savana Ltd. of Sri Lanka have organised the small farmers in the villages into small groups of 5-7 members belonging to different families living in the vicinity of each other and whose income are of the same level. Such groups were clustered at village level. These village clusters are brought together into an Association working like a Producer Co-operative. 83 ha of organic tea is planted by these groups. NASAA, Australia is helping them with the inspection. In 1999, the 151 tea growers of Gemi Seva Sevana produced 6,937 kg of organic tea, of which 1,450 kg was exported. Their produce is sold mainly to Switzerland with the help of the Swiss Association for International Cooperation (http://forests.org).

Bio Foods started as a registered company in 1990 in Sri Lanka. Its main areas of business are organic cultivation, harvesting and processing quality organic green teas, organic black teas, naturally flavoured and herbal teas, spices and coffee. They also have an extension and local (internal) inspection section, produce collecting centres, processing factories, an export and an administration section. The farmers are organized at village level in farmer committes. Small growers of Small Organic Farmers' Association (SOFA) under Bio Foods have cultivated 214 ha organic tea in Kandy certified by Naturland. The small growers are getting higher guaranteed price for green leaf from Bio Foods. Advisory services to the growers are provided and cattle are donated to farmer families to assist in preparation of compost manure and to generate additional income from milk sales. This company also purchases green leaf from the small growers. All products are certified organic, by SKAL, Naturland, IMO or JAS, depending on the country of import and the requirements of the buyer. Certification costs are paid for by Bio Foods and the certificates are owned by them. However, Bio Foods is reimbursed by the Small Organic Farmers' Association for certification costs for those farmers who are member of SOFA. The tea producers of Bio Foods are also registered by FLO (Fairtrade Labelling Organizations) International and part of the tea is sold under fair-trade conditions, in which case the buyer has a fair-trade licence from an FLO member[1].

Box 3.1

Fair Trade with Small Farmers

Fair trade organizations play an important role in distributing a number of product groups from developing countries especially bananas, cocoa, coffee, honey and tea, in Northern Europe. The primary concern of the fair trade movement is the social and economic situation of farmers and producers in developing countries rather than organic agriculture as an ideal. However the two movements are converging in many aspects, as both contribute to environmental sound and sustainable development.

Fair Trade Criteria

Fair price

Prices paid for tea should at least cover the costs of production and provide for an extra margin for the improvement of working and living conditions and for future investments. These might include conversion to organic cultivation, education and training for small farmers.

Long-term relationship

Producers are very vulnerable to the erratic movements of the international tea market. Fair Trade organisations, in principle, maintain trade relationships on a long-term basis which gives producers security. This allows them to plan social development projects or achieve long-term goals such as transition to organic cultivation.

Advance payment

At the producers' request, part of the product value can be paid in the form of an advance payment. This prevents producers from falling into debt before they harvest their crop. Small growers can use this money, for example, to improve the productivity of their land and therefore the quality of their tea.

Packing

Tea should preferably be packed in the producer country, so that producers receive the added value and new jobs are created. This is particularly Important for producers who want to be more than the mere suppliers of a commodity which is destined to be refined, blended and packed elsewhere.

Source: http://www.eftafairtrade.org/

Tea is a unique fair trade product in that there is no minimum fair trade price yet established. India, Sri Lanka and Tanzania together supply 85 percent of all fair trade teas. There are around 27 exporters and 49 importers involved in the trade of fair trade teas. Some well known fair trade suppliers are Tea Promoters of Ibdia and Stassen from Sri Lanka. The market for fair trade tea is dominated by the UK, followed by Germany and The Netherlands. The recent introduction of fair trade tea in the US seems to be quite successful (Intergovernmental Group on Tea, FAO, document: CCP: TE 03/4). This tea in the UK market mainly imported by e.g. Clipper and Traidcraft, Intergovernmental Group on Tea, FAO, Rome in its meeting held in Colombo, Sri Lanka during 18-20 August, 2003 reported that Fair trade premium of Euro 0.50 or Euro 1.00 per kilogram of tea could be attractive. Although fair trade and organic tea sales grew by 15-20 percent a year, market volumes remained small at 1266 tones per year in 2002, or 0.3 percent of the total combined tea market of Western Europe, North America and Japan. The Group also concluded that the slow progress in building the fair trade market suggested that this would remain a small niche market and would not solve the structural imbalance in the world tea market, although it might offer a lucrative outlet for certain producers (Intergovernmental Group on Tea, FAO, document: CCP: TE 03/7).

However, the expansion of organic tea production by smallholders has been restricted by factors such as substantial yield drop during and after the conversion; changing attitude of farmers (farmers who had converted their lands for organic production have later given up their commitment); high labour requirements especially for compost making; poor access to capital and credit; and difficulty in rearing livestock, which in turn provides manure for organic crop production. If the farmer's income could be increased by higher productivity, it would encourage their involvement in organic tea production.

China is the ancient birthplace of tea. One of the most important Chinese production centres is located to the South of Beijing, along the Pacific coast. Around 50 tea gardens of this region are currently in conversion or fully converted (www.sippo.ch). The farmers are very poor to afford pesticides and other chemical inputs for their gardens and produce chemical free teas. China has many favourable conditions to develop organic tea, including pollution-free environment in some mountainous and remote tea areas and supportive policies by some local government. In China, first tea estate was converted to organic estate in 1990s. The production of organic tea increased more than 166 percent in 2004 compared with that of 2000. Till 2003, the organic tea plantation spread from Zhejiang to Yunan, Hubei, Hunan, Jiangxi, Anhui and then Fujian province (Table 3.4). China's first batch of organic teas was produced in Lin'an country in Zhejiang in June 1990 was certified by SKAL and exported to Holland (http://www.itmonline.org.arts/organics). Wuyi country of Zhejiang province developed organic tea around 1000 ha since 1995 and known as 'home of Chinese organic tea'. 16 enterprises of Wuyi country were certified for their produce. Sichuan started later than provinces such as Anhui and Jiangxi and now boasts seven gardens

Table 3.4. Estimated organic tea area in some provinces in China.

Province	Year	Area (ha)*
Zhejiang	2005	6666.7
Hubei	2005	3000.0
Yunan	2005	1330.0
Hunan	2005	3300.0
Fujian	2006	2000.0

* Nationally and internationally certified organic tea area is 18,180 ha which 6.1% is over land area under organic management; Source: Ref. 3.

near Le Shan producing 50 mt of organic teas[2]. Fujian Xian Yang Yang Food & Technology Co., Ltd., Ningde City, Fujian Province, China has tea cultivation plot of about 335 hectares, including a 67 hectare organic tea garden. Organic tea leaf is widely used to produce fresh concentrated tea juice (Source: http://xianyy.en.alibaba.com/aboutus.html). Currently, there are more than 300 certified organic tea producers in China. During 2003, a guideline issued by the Agriculture Ministry of the People's Republic of China that all certification body must obtain approval from the certification and accreditation regulatory department of the State Council. The production of organic tea attained 12,000 tons accounting 0.5 percent of total Chinese tea production (Table 3.5). It is estimated that the volume of production and export of organic tea in China would attain 14,000 and 5,000 metric tons respectively in 2005 (Table 3.6). The villagers of the hilly area of Xiuning country under Anhui province are cultivating organic tea and selling their product in Beijing and other cities. The export price of certified organic tea is currently 50 percent higher than the price of the conventional tea. It has been reported that the income per person was calculated 10.7 times more (2600 Yuan) during 2003 than the income of five years back (242 Yuan) by cultivating organic tea [3]. A small, but growing number of poor farmers of Shifo village, in Baojia Township, Yuexi country adopted organic cultivation. In 1997, a group of people of Shifo converted their conventional tea areas into organic ones. They were promished education and training by OFDC and the German Governmental aid

Table 3.5. Land area under organic management and percent organic tea area etc of some organic tea-producing countries.

Countries	Land area under organic management (A)	Area under organic tea* (ha)	Percent organic tea area over (A)	Percent organic tea over convensional production
India	76326	4500	5.89	0.4
China	298990	6060	2.0	0.5
Japan	29151	2500	8.58	1.8
Sri Lanka	15215	1300	8.54	0.3
Tanzania	55867	100	0.33	0.2
Malaysia	600	20	3.33	0.1

* Certified by the International certification agencies;

Table 3.6. Production, export and domestic consumption of organic tea in China.

Year	Production* (Mt)	Export (Mt)	Domestic Consumption (Mt)	Export price (US$/Mt)		Domestic (US$/Mt)		Revenue growth (%)
				Conventional	Organic	Conventional	Organic	
1996	100	100		2000	4000			
1997	200	200		1990	4000			100
1998	1500	1000	500	1980	3900	4000	5000	700
1999	3000	1500	1500	1970	3800	4000	5000	106
2000	4500	2500	2000	1970	3800	4000	5000	48
2001	6000	3000	3000	1970	3800	4000	5000	35
2002	8000	3500	4500	1970	3800	4000	5000	36
2003	10000	4000	6000	1980	3900	4000	5000	27
2004	12000	4500	7500	1980	3900	4000	5000	21
2005**	14000	5000	9000	1980	3900	4000	5000	17

* National and internationally certified; ** Estimated; Source: Ref. 3.

Organization (GTZ) and enticed by the prospect of premium market prices. Later in 2000, the Shifo farmers formed the 'Organic Tea Growers' Association' in China. Around 10 ha (157 mu) organic tea was certified as organic by Organic Food Development Centre (OFDC) in 2001 and the association obtained their trademark 'Shifosi tea' during 2002. There was another 21 ha (equivalent to 340 mu) put under conversion in 2003 and total tea area will be converted and ready for organic certification by 2006 [4].

Zhejiang Caiyunjian Tea Industry Co., Ltd. located in Jinhua city and established in 1992 is one of the largest professional manufactures of organic tea in China, involving tea growing, processing and marketing. They have over 800 hectares of tea plantation and 8 processing factories. The annual production of made tea is over 2,000 tons. Some of their organic tea plantations are Jin'shan tea plantation, Tai'shan tea plantation, Jiu'feng'shan tea plantation, Da'yuan'tang tea plantation, Ma'kou tea plantation and Yan'tou'shan tea plantation (Fig. 3.5). Caiyunjian became a member of IFOAM in 2000 and the company has been certified as organic tea producer by IMO, NOP, JONA, OTRDC as well as OFDC. And its organic teas have been exported far to USA, Japan, Germany and other tea-consuming countries. Caiyunjian's main green tea products including Jinshan Green Sword, Double Dragon Green Sword, Newtoop Mist Green, Taishan Buddha's Eyebrow, High Mountain Lungching Tea, Steamed Green tea (Sencha), and other tea, such as black tea, jasmine tea, Kuding tea, oolong tea and chrysanthemum tea.

Da Zhang Shan organic tea garden located in the Jiangxi province of China produces a large variety of green and black teas including Chun Mee, gunpowder, jasmine tea, congou black tea, CTC black tea, green fannings and black fannings. This tea garden is certified by BCS OKO-Granite.

The Organic Tea Research and Development Center (OTRDC), under the Tea Research Institute of the China Academy of Agricultural Sciences (CAAS), is a domestic certification body specializing in organic teas, beverages, and herb products. OTRDC has developed the standards based on the IFOAM basic standards for organic tea, which is known as Zhejiang provincial standards and has now become national standard. OTRDC supervises proper use of organic logo and to provide training, and also to develop the standards of organic tea. Since its establishment in March 1999, it has issued certificates to a total of 110 organic tea gardens. OTRDC certified products are widely accepted by Chinese consumers and spread quickly in domestic market. It has begun promoting its certification to the world market. Despite ORTDC, others are Tea Research Institute of the Chinese Academy of Agricultural Sciences, OFDC and COPCC. China Green Food Development Center (CGFDC), in Beijing, was the first organization in organic/green food development in China. It is a national level institution with the authority to certify agricultural products as being free of chemical contamination. This center was officially established in November 1992 under the jurisdiction of the Ministry of Agriculture. Further, Jiangsu Ruikang Organic Food Trade (JROFT) was established in 1995 and is one of the pioneers of the organic agricultural movement in China. JROFT cultivating organic tea, organic licorice, and organic star anise and conform to the EEC (European Economic Community) regulations and the standards of OFDC for organic agriculture. Foreign organisations that join certification process actively include IMO, BCS (Bio Control Systems) and ECOCERT, OCIA (Organic Crop Improvement Association) in the U.S, UK's Soil Association and JONA etc. ECOCERT and BCS have established local branches as joint ventures with Chinese partners. In China, the organic tea production and certification are mainly restricted among three groups viz. companies, corporation and small holders. The first group directly managed cultivation, processing and selling of organic tea. The quality of tea produced under companies is also stable. The second group is Trade Corporation which is a main part of the associations and some tea planters having large scale plantation. This group can easily broaden the area as well as production of organic tea. The small holders are supported by the government.

There are about 2,500 hectares of organic tea fields in Japan that yielded around 3,500 tons of green tea. Very small requirement of black tea is imported mainly from India and Sri Lanka. There are number of organic tea plantations in Japan. For example, Hou sui-en located in Mie prefecture, Tsuen's organic tea farms (Fig. 3.6) located in Watsuka cho, near Uji city in Kyoto producing green tea. Japanese tea planters grow organic green tea (Fig. 3.7). Tsuen's farms have started cultivation of organic tea since 1987 and certified by Jona (Japan Organic & Natural Foods).

In Nepal, total production of organic tea is 60 tons [5]. Guranse tea estate, one of two organic tea estates, covers 250 hectares in Dhankuta situated in the eastern hills of Nepal at an altitude of 3,300 feet to 7,300 feet, one of the highest-altitude tea estates in the world. Guranse (which means "rhododendron" in Nepali) was certified as organic in 2000 by the NASAA, Australia. Guranse Tea Estate has obtained ISO 9001:2000 Certificate from TUV Rheinland, Germany. The estate is planted with young clonal bushes. Despite Guranse tea estate, Kanchanjangha tea estate (KTE) is another organic tea estate located at Ranitar in the remote hilly region of Panchthar district in eastern Nepal bordering the tea area of Darjeeling at an altitude of 1300 - 1800 m. The conversion to organic started in 1995 and certified the garden as organic from June 1, 1997. KTE is one of the finest orthodox tea-producing gardens in Nepal that grows purely organic black and green tea which is certified by National Association for Sustainable Agriculture, Australia (NASAA) accredited to IFOAM according to the rule of the European Union. KTE exports 95 percent of its produce to Europe and Japan directly. It also produces tea bags branded as 'Nepal Organic Tea Bag' that are exported and supplied locally to the renowned hotels and resorts in Nepal. Because of its relatively low cost of production, Nepalese organic tea is sold at much lower price. Shangrila Agro Trading Pvt. Ltd., Manbhawan, Lalitpur, Nepal have taken the task of promoting organic tea of KTE in the domestic and international markets. Recently, the Nepal Tree Crop Global Development Alliance (NTC-GDA) of Winrock International is promoting Nepal tea in Germany and USA. Winrock has been helping

Tai'shan tea plantation

Jin'shan tea plantation

Yan'toushan tea plantation

Jiu'feng'shan tea plantation

Ma'kou tea plantation

Da'yuan'tang tea plantation

Fig. 3.5. Some organic tea estates in China. Source: http://www.chinaorganictea.com

Fig. 3.6. Tsuen's organic tea farm in Watsuka cho, near Uji city in Kyoto, Japan. Source: http://www.tsuentea.com / engorganic.htm

Fig. 3.7. Plucking in Japanese organic tea plantation. (enlarge view)

Nepal tea producers to commercially penetrate the US speciality tea including organic tea market. The discussion for making partnership with TAZO tea, the main supplier of specialty tea to the Starbucks chain is in process.

In Tanzania, Mufindi Tea Company Ltd., produces organic tea at its Luponde tea estate and is located in the Livingstonia Mountains at 6,800 to 7,200 feet above sea level, on the edge of the Great Rift Valley near Lake Nyasa. The large, wiry, dark green leaves brew to a pale green cup, with a sea breeze, lemony aroma. This tea is delicate, slightly sweet and has a wonderful balanced body. This tea estate certified by ECOCERT. In fact, tea grown here was organic by default. The company began organic tea production in 1989. Herkulu tea estate is located in the Usambara region of Tanzania covers 230 hectares. It has been certified according to EEC Bio-regulation 2092/91 by IMO.

The development of organic farming in Taiwan was initiated in 1988. The farm land dedicated to organic culture in Taiwan accounts for only 0.3 percent of the total agricultural area, indicating there is great potential for the development of organic farming in Taiwan. There are several research institutes and universities exerting efforts on the improvement of organic farming in Taiwan, and National Chung-Hsing University (NCHU) plays a leading role in the development of organic agriculture in this area. The Taiwanese government has selected five non-governmental institutions to oversee certification process of organic products. The institutions include Mokichi Okada International Association (MOA), Tzu-hsin Organic Agriculture Foundation (TOAF), Taiwan Organic Production Association (TOPA), Chinese Organic Agriculture Association (COAA) and Formosa Organic Agriculture Association (FOAA).

Certification of all organic food is mandatory by one of the five institutions prior to be sold to consumers. Currently in Taiwan only 80 ha is under organic tea. MOA, TOPA and TOAF certify about 54 organic tea growers with 57.8 ha of organic tea gardens. Most organic tea produced is partially fermented tea, oolong tea. Only paochung tea is produced from 10 ha organic gardens. The organic tea areas are mostly located at Nantou country with 29 growers in 2003. Organic tea is also grown at 4600 feet in the broadleaf forest zone that contains the villages of Tungpu and Meishan. The medium rolled olive green leaves have a sweet aroma and open into dark green, extremely large whole leaves. The cup is light green with a floral aroma and a full body not generally found in an oolong. About 66 percent organic tea producers export their tea directly to Japan, USA, Germany and other EU countries through posts. Organic tea price range from 20 US$/lb to 260 US$/lb (equivalent to 50 to 650 US$/kg) [6].

In Malaysia, around 20 ha are under organic tea and 6000 kg tea is produced. It is exported to USA, Australia, Japan and Taiwan. (Personal communication with Mr. Alvin Heng, Sabah Tea, Malaysia; 14 January, 2005). Sabah Tea has a tea plantation in the rainforests of Borneo - 2,272 feet above sea level made its mark by becoming Malaysia's first organically certified tea. The plantation is certified for organic production and processing by SKAL International B.V of the Netherlands.

Kazi and Kazi Tea Estate, Tetulia in Panchagar district is the only organic tea estate at present in Bangladesh which has been certified by SGS during November, 2004. Tea is sold in Chittagong auction. The estimated production is 1.5 mt in 2005 (personal communication with Dr. Monjur Hossain, PMTC, Bangladesh; 19 April, 2005). The company is looking to launch a small quantity of tea in packeted loose leaf and tea bags, which will be sold through their affiliated organic outlet Meena Bazar in Dhaka city, and other outlets or retailers. Although the company is not making any direct exports yet, they were able to sell their tea at Harrods in London in 2004.

Vietnam organic tea production is just at the initial stage. Recently, Thai Nguyen University of Agriculture and Forestry carried out one pilot project on organic tea marketing and development in Thai Nguyen province in Northen zone of Vietnam (70 km North of Hanoi) with the financial support from the Government and NGO. This project has already been completed in 2003 and the objectives of the project were to improve quality, increase range of tea products, improve marketing, get certification and develop a system for external quality control. (Personal communication with Dr. Minh Dang, Thai Nguyen University of Agriculture and Forestry, Dept. Soil Science, Vietnam; 13 March, 2005 and at http://www.cidsevietnam.org/who. htm).

In Papua New Guinea, Garaina tea estate located in the Morobe Province was established in 1950 with seeds originated from India. This estate was certified by NASSA, Australia which is accredited by IFOAM. Garaina is a remote rural area and there are no roads, accessibility is only by air. The tea is flown out by small single engine planes to Wau, connecting to Lae then arrives in the National Capital Port Moresby for delivery to the buyers.

The organic tea production in Tanzania, Taiwan, Malaysia,

Vietnam, Japan, Argentina and Indonesia are very marginal and do not have a significant influence on the world market for organic tea. Except for Japan and China, other organic tea-producing countries do not yet have a domestic market, which will be needed to build up a sustainable business.

Share of organic and total tea production

This section provides an overview of the pattern of organic tea production relative to total (organic and conventional) tea production in some organic tea-producing countries. By 'organic production' it means the volume of teas that is harvested and then processed at tea estate level and which is certified as organic. When comparing levels of the organic share with the levels of the total production, it should be taken into consideration that organic is less intensive than conventional agriculture and organic yields are much lower because of this. Furthermore, organic way of tea production is mainly concentrated in few countries and that too in the initial stage. Therefore, most of the organic tea production data presented as a share of total (organic and conventional) production is much lower (Fig. 3.8). Comparatively, the organic tea production as a share of total production in Japan is found highest among the organic tea producing countries. Malaysia had the lowest. China, India and Sri

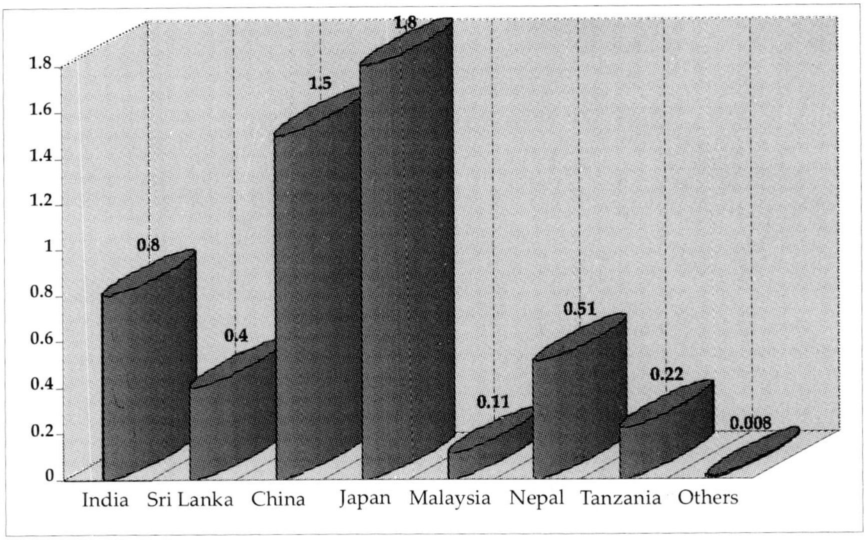

Fig. 3.8. Organic tea production as a percentage of total tea production in some important organic tea producing countries in 2004.

Lanka though produce higher quantity of conventional tea but organic tea production when considered as a share of total production had Japan with the highest.

References

1. Dankers, C. and Liu, P. 2003. Environment and Social Standards, Certification and Labeling for Cash Crops. FAO, Rome, Italy. p. 103.
2. Kung, L. 2005. Exploring speciality teas in Southwest China. *Tea & Coffee Trade J.* **177**(2): 60-65.
3. Liu xin, Shangwen, F., Zhangyu, Amin, S. and Zhichang, W. 2004. The development of organic tea in Chinese mainland. *Proc. Intl. Conf. on O-CHA (tea) Culture and Sci.*, Nov. 4-6, 2004 Shizuoka, Japan, pp. 732-735.
4. Sanders, R. 2004. The anatomy of an organic village. *China Review* **28**: 1-4.
5. Aryal, M. 2005. A new leaf. *Nepal News* (http://www.nepalitimes.com) **261**: 1-3.
6. Lin, M-L. and Chen, T. C. 2004. Tea cultivation in Taiwan. *Proc. Intl. Conf. on O- CHA (tea) Culture and Sci.*, Nov. 4-6, 2004 Shizuoka, Japan, pp. 820-823.

Chapter 4

Botany and Propagation

Botany

More than 3,000 different teas are produced in many places around the world and they are all made from one plant – the *Camellia sinensis* (L.) O. Kuntze [1]. This is normally a diploid ($2n$ = 30 chromosomes) but polyploids occur. Various types within the tea population are often mentioned. Two that are well-known are the China and Assam and Cambod is the less common. These are often referred to as *Camellia sinensis*, *C. assamica* and *C. assamica* subsp *lasiocalyx* respectively [2]. Alternatively the China type may be described as *C. sinensis* var. *sinensis* (L.) and the Assam type are *C. sinensis* var. *assamica*. Mohanan and Sharma [3] have listed the characteristics of each of these taxa. In addition to the three principal tea-producing taxa, there are a number of species with similar characteristics which do not produce the necessary chemicals but may be involved in hybridization. Most important are *C. irrawadiensis* Barua (Wilson's *Camellia)* and *C. taliensis* (W. W. Smith) Melchoir (Forest's *Camellia*). A few other taxonomically important non-tea *camellia* spp. are *C. japonica* L., *C. kissi* Wall, *C. pitardii* var. *yunnanica* Cohen-Stuart, *C. sasanqua* Thunb., *C. reticulata* Lindl., C. caudata Wall. These species of *Camellia* seem to have been involved in the evolution of the present-day commercial tea population.

The tea varieties have developed into a complex taxonomic group of related plants. Tea is essentially an outbreeding crop and its races are highly infertile. As a result of enormous natural hybridization not only between the tea species but also with *Camellias* and large-scale dispersal of tea seeds in South-East Asia over the centuries has resulted in extreme heterogeneity in the existing seedling population.

Botanical composition of *jats*

Distinctive types of tea based on foliar characters and the district or the garden growing the seed is given to the strain and this is known as the *'jat'* of the tea. The term *'jat'* in tea nomenclature is considered close to 'variety'. The multi-stemmed China type has two *jat* - *'Low-jat'* and *'High jat'*. The former *'jat'* has narrow leaves, profusely indented margin and flat, rather dull-coloured laminae, whereas the latter has comparatively larger leaves, wider in proportion with blunter tips, fewer serrations and a crinkled surface [4]. The single-stemmed Assam type has *Burma, Lushai, Manipuri, light-leaved Assam* and the *dark-leaved Assam jats. Burma jat* comes from Myanmar hills and has spiny, crinkled and slightly erect leaves and has dark green, older leaves having a bluish tinge.

Biology of the plant

The tea plant under natural conditions is an evergreen tree and wildly grows into a medium-size tree. Under cultivation, it is pruned down and trained as low-spreading bush to ensure that a maximum crop of young shoots can be obtained and maintained. The root system of a tea bush can be divided into main roots, subsidiary roots and feeder roots. The lateral roots give rise to a surface mat of feeding roots which lack root hairs when mature. Root system varies in tea bushes depending on genetic make-up and soil environment. Starch is stored in roots.

Tea flowers appear in the axils of scale leaves, either solitary or in clusters, with short pedicels, fragrant; calyx persistent with 5 - 7 sepals; petals 5 - 7, white, waxy, obovate, concave; stamens numerous with 2-celled yellow anthers; petal and outer stamens united for a short distance at base; ovary superior, hairy, with 4 - 6 ovules per carpel; styles short, free to almost completely fused, stigmatic lobs 3-5.

Fruits are capsule thick walled, brownish green, 3-lobed and usually 3-celled, becoming slightly rough at maturity, dehiscing during the dry period by splitting from apex into 3 valves.

Propagation

The traditional source of planting material is seed. Though it is convenient to use, a seed population is composed of a large number of genetically heterozygous genotypes. However, it can be fitted into a wide range of environmental and field conditions without much change in its overall performance. In order to produce a uniform crop with predetermined characters, vegetative propagation of superior clones are practiced. Vegetative propagation through single-node leaf cuttings is reliable and a economic method. Cuttings of the tea estate selection, clonal cuttings, bi-clonal seedlings and polyclonal seeds are also considered as planting materials. Low input responsive clones or seed stocks having wide adaptability with tolerant to different stresses should be preferred in the organic tea estates. Genetically modified seeds or planting materials are not used in the nursery as their use is prohibited in the organic field.

Location of nursery

It is recommended to establish own nurseries in the tea garden, in order to ensure a continuous supply of untreated and healthy plants. The nursery should preferably be located between low and mid-elevations and also at a convenient place where constant supervision is possible. It would be helpful if the site is located near a perennial water source and is desirable if it is naturally protected from wind. In fact, the area should have a shelter belt planted with evergreen trees like *Acacia auriculiformis, Cassia siamea, Melia azedarach* (Ghora neem) etc to create a favourable microclimate to protect the nursery from dry wind. Planting of *Crotalaria* ssp., *Tephrosia candida, Sesbania* spp., would be useful as natural shade. The slope of the land is also important. Flat land certainly needs drain. Since water-logging leads to several problems in the nursery therefore a proper drainage at the site is essential. It is convenient to situate the nursery not too distant from the area to be planted.

Seed production

Selected plants are planted in 5 - 6 m apart in the seed orchards (known as '*baries*' in the industry). The land of the seed *bari* should be flat or with a gentle slope with good drainage. The plant should be allowed to grow freely, apart from hygienic removal of dead branches with occasional thinning of branches to ensure entry of adequate light through canopy; trees must not interlock. A ground cover should be grown for the first few years and must be kept clear and free from weeds when production of seed starts. Seeds are borne mostly on the surface of canopy of the tree and collected on the ground. Tea seeds if kept open lose their viability rapidly with time. Generally, a kilogram of seeds will have 275 to 450 seeds depending upon their size. The seeds should be sorted by floatation immediately before sowing them in the nursery. All those which have sunk within 24 h are the best.

All seeds and plant materials must be of organic origin. Since organic tea cultivated area at present is very limited therefore availability of such seeds and plant materials from certified organic area is difficult. Till such time, seeds and plant materials from conventional tea estates, not treated with any chemical, can be used. The nursery for raising organic plots should be clearly separated from conventional nursery if both the activities are being carried out in the same estate.

The sleeve size should be around 15 cm x 25 cm in the nursery. Low energy and plant base inputs approved for organic agriculture should only be used for soil treatments, nutrient managements and pest management. Ghosh Hajra [5] has discussed the vegetative and seed propagation techniques in detail.

References

1. Pettigrew, J. 1999. *Tea*. PRC Publishing Ltd., London, p. 96.
2. Wight, W. 1962. Tea classification revised. *Current Sci.* **31**: 298 - 299.
3. Mohonan, M. and Sharma, V. S. 1981. Morphology and systematics of some tea (*Camellia* spp.) cultivars. *Proc. PLACROSYM IV*, 3-5 Dec., 1981, Mysore, India pp. 391-400.
4. Subramaniam, B. 1995. *Tea in India*. Publication & Information Directorate and Willey Eastern Limited, New Delhi.
5. Ghosh Hajra, N. 2001. *Tea Cultivation: Comprehensive Treatise*. Int. Book Distribution Co., Lucknow, p.518.

Chapter 5

Climatic Limitations, Soil Requirements and Management

Climate

The tea plant prefers warm temperatures, humidity and plenty of rainfall. It is grown well in areas where annual rainfall varies from 1150 to 6000 mm. Tea should not normally be grown in areas where the rainfall is below 1,150 mm, unless irrigation is available. The effect of precipitation is perhaps more manifested by its influence on moisture status of the soil and in inducing vegetative growth. Therefore, distribution of rainfall over the year is as vital as the total annual rainfall. In Sri Lanka and South India where South West monsoon is heavy, excessive rain water logs the soil and slows down the growth of tea. When soil is saturated particularly at the peak of the rainy season, some rainfall is lost by runoff (for example Darjeeling) until the ground cover has reached at least 60 per cent [1]. The adverse effect of either excess or lack of adequate rainfall can be compensated by drainage or irrigation respectively.

Climate influences yield, crop distribution and quality. Temperature affects tea yield by influencing rate of photosynthesis and controlling growth and dormancy. In general, the ambient temperature within 13°C and 28-32°C is conducive for growth the tea. In many instances soil temperature is of greater importance to plant life than air temperature and soil temperature influences the growth and yields of tea. Carr and coworker [2] concluded that soil temperature is an important variable, with a lower limit of about 20°C, at 0.3 m under short grass surface (or 16°C beneath a canopy tea), below which shoot extension rates will be reduced. The corresponding upper levels are 29°C and 25°C respectively.

Relative humidity of 80 - 90 percent is favourable during the growth period of tea plants, below 50 percent, shoot growth is inhibited and below 40 per cent growth is adversely affected.

Solar radiation is the source of energy which sustains organic life on the earth surface. Crop production is in fact an exploitation of solar radiation. The intensity and duration of sunshine has also an influence on the growth of tea plant. The hours of bright sunshine are fewer in the rainy season of Darjeling hills, Bangadesh and other places due to the overcast clouds.

Day length influences growth and dormancy in tea bushes. When days of less than 11hr 15 min duration last for at least six weeks tea bushes become dormant.

The China variety in particular and the hybrids in general produce very fine, flavoury tea when they are grown above an altitude of about 1,100 m as in the case of Darjeeling and in the high-grown areas of South India and Sri Lanka. Surprisingly, such flavour does not develop at high altitudes in Java of Indonesia and East Africa. This probably is because of the uniform weather conditions that prevail resulting in harvesting of crop throughout the year [3]. China hybrid bushes from Darjeeling fails to reproduce the full hill muscatel flavour if grown in the plains of Assam and Dooars. It may thus be stated that the Darjeeling teas owe their unique flavour partly to the type of bush and partly to the climate. Altitudes affects shoot growth rate and influences the quality of processed tea [4].

Wind causes injury by toppling and uprooting plants, breaking stems and branches, causing stem malformation, and injuring leaves. In addition, wind erodes soil inhibit growth by adversely affecting physiological processes, and increases risk of insect attack, disease and fire. The velocity of wind increases with elevation. Tea plantations in windy regions should also be protected by wind breakers to reduce the intensity of evapo-transpiration. For boundary/estate barrier, tall trees viz. *Casuarina montana, Eucalyptus globulus* Labill., *E. robusta, E. saligna* Sm., *Grevillea robusta* A Cunn. and *Dalbergia sisso Roxb. ex DC.* at 4.5 x 4.5 m spacing, then the medium trees such as *Cassia siame* Lam., *Acacia auriculiformis* A. Cunn. ex Benth., *Erythrina poeppigiana* (Walpers) O. F. Cook, *Melia azadirach* L., *Pongamia glabra* Vent. at 4 x 4 m spacing and dwarf trees viz. *Acacia elata* A. Cunn. ex Benth., *A. mearnsii* De Wild., *Murraya exotica* L., *M. koenigii* (L.) Spreng., *Hibiscus roseus, Dohonea viscose* L. and *Cypruss* sp. can be planted at 3 x 3 m spacing. Lines of wind belt trees should be at a right angle to the direction of the wind. Smaller trees/shrubs viz. *Leucaenea leucocephala, vitex nugundo* and *Sesbania* spp. (Daincha), fodder plants and leguminous plants can also be planted at intervals of 1 metre. The border trees and shrubs take at least 2-3 years to yield adequate biomass. The border trees also add to the biological diversity and improve the overall eco balance.

Hail is unpredictable and of a localized nature. Hail in tea-growing districts viz. Cachar, Dooars, Darjeeling, and Terai of North-Eastern India are almost a regular phenomenon every year during spring and cause severe losses. Hail storms occur in Darjeeling hills in one valley or other during March to May, but the end of March to early April is the period when it occurs the most. Other tea areas, especially in Kericho and Nandi Hills of Kenya, Bangladesh and China are prone to hail storms. Hail damage areas are susceptible to mites, thrips and blister blight. All possible care should be taken to keep them free from pests and blight. All hail-damaged areas should as quick as possible be sprayed with copper fungicide mixed with spreading agent. Young tea and pruned section must have priority over other sections. Plucking should be done very liberally on these areas.

Soil

Tea is grown in a variety of soil types that are widely different in origin and geographical formation. These soil types have developed from diverse parent rock material under different climatic conditions. Tea grows best on well-drained fertile soil on high lands having a good depth, acidic pH in the range 4.5 to 5.5 and organic matter status should be medium to high level (more than 2 percent). It is better to avoid planting of tea in soils with pH value higher than 5.5. The soil pH should be maintained at about 5.0 by application of agricultural lime or dolomite lime once in a pruning cycle. The quantity of liming material should be determined on the basis of soil pH, rainfall and crop yield. Dolomite also serves as a source of magnesium.

Shallow and compacted sub-soils limit root growth and plants are liable to suffer from draught during dry period and water-logging during the rains. Any hard pan or concretions in the subsoil within 2 m depths is not desirable. The depth of ground water table should not be less than 90 cm for good growth of tea. Catchment planning is required for improved soil and water management practices in a tea estate for which land survey designed to identify all major and minor topographical features needs to be carried out.

Constant soil testing is mandated, and even when the land becomes free of contaminants and new planting is carried out, the soil must be constantly monitored to achieve the level for which certification can be granted.

Box 5.1

Soil Performances

An important objective in organic farming systems is the achievement and sustenance of the soil's fertility. The fertility of a soil can be measured by the variety of species that grow in it. Soil is a decisive agricultural production factor which, when fertile, can fulfil a variety of purposes:

- Growth medium for plants
- Water storage and supply for plants
- Decomposition for organic material
- Anti-phytopathogenic potential (suppression of soil-borne diseases)
- Nutrient reservoir and availability

Organic matter

Soil organic matter refers to the organic fraction of the soil; it includes plant and animal residues at various stages of decomposition, cells (living and dead) and tissues of microbes, and substances synthesized by the soil population. In a fertile soil, the function of organic matter is both direct and indirect. Its direct role is concerned with the provision of plant nutrients via the process of decomposition and mineralization, its indirect role is associated with its effect on the physiochemical properties of the soil. The soil organic matter has profound influence on the fertility status of a soil and the comparative ratings are presented below:

Percentage of organic carbon	Fertility status
Below 0.6	Very low, unsuitable
0.6 - 0.8	Moderate, marginally satisfactory
Above 0.8	Satisfactory, suitable for tea cultivation

Organic matter serves as the principal store house for anions such as nitrates, sulphates, borates, molybdates, and chlorides that are essential for plant growth. The organic matter content predominantly contributes for high Cation Exchange Capacity values which influences the nutrient fixation and availability. It acts as a buffer against rapid changes caused by acidity, alkalinity, salinity, pesticides and toxic heavy metals. Organic matter also supplies food for beneficial soil organisms like earthworms, symbiotic nitrogen-fixing bacteria, and micorrhizae (beneficial fungi).

An increase and better use of organic 'waste' and green manures (animal manure, crop residues, household refuse and leguminous plants collected within and outside the plantation) as organic fertilizer would greatly enhance nutrient availability and the biological functioning of the soil. The soil would be softened and water retension improved. It would also make plants more resistant to pests and diseases and prevent 'soil acidification'[5]. Tea soil vary in their organic matter contents ranging from less than 1 percent in tropical, high leached soil to over 25 percent in peat (bheel) soil. However, the tea plant generates a lot of organic matter from leaf-fall and recycling of pruning litters in the well-maintained garden (Table 5.1). Tea bushes

Table 5.1. Dry matter production andnutrient content of plant parts formed above pruning cuts of bushes (clone B/6/60) under regular plucking.

Plant part above Pruning cut	Cumulative dry matter (tonnes ha^{-1})	Nutrient removal (kg ha^{-1})		
		N	P	K
Up to first year				
Crop	1.20	48	6	24
Foliage	3.30	122	15	60
Twigs	0.82	13	2	10
Wood	1.79	41	8	15
Total	7.11	224	31	109
Up to second year				
Crop	4.80	192	18	96
Foliage	4.74	164	12	46
Twig	3.50	64	7	24
Wood	7.02	68	10	76
Leaf litter	0.60	13	2	11
Total	20.66	601	49	202
Up to third year				
Crop	10.80	414	46	201
Foliage	7.33	249	30	81
Twig	6.74	142	25	81
Wood	17.27	162	45	64
Leaf litter	1.80	61	10	7
Total	43.94	1028	166	412

Source: UPASI, Tea Research Foundation, India

Note: Values are based on studies carried out in conventional tea plantations. Therefore, values can only be approximately used for the situations on organic tea plantations.

are pruned to keep them at a manageable plucking height. The pruning litters are cut into small pieces and then shredded. For decomposition, fungal cultures of *Trichoderma viridae*, *Trichurus spiralis* and *Paeciliomyces fusiporus* can be added @ 300 gm/metric tons of material and 100 percent moisture level should be maintained at the initial period for best result.

Managing tea soil health

Good soil health is the precondition for good plant health. The maintenance of good soil structure and fertility through proper husbandry practices are of vital management considerations for tea soils in the organic tea estates. Unplanned land use, bad soil structure, inadequate drainage, soil compaction etc reflects decline of productive potential of the tea soils. These adverse conditions can be checked by adopting proper soil management practices. These are:

Water conservation

Water may enter a plantation as rain, flood or irrigation of water, groundwater or air humidity. In general, 93-95 percent of the total annual rainfall received during the monsoon sometimes cause waterlogging and drainage problem in many tea estates of Assam and other parts of India. The rest (5-7 percent) falls during the period from November to April which sometimes leads high soil moisture deficit. Thus, tea may suffer from excess and deficit of the soil moisture in the same year. In marginal rainfall areas, every drop of rain is important for growth of the tea plant. Planters can influence the availability of water in the soil by improving soil structure and storage capacity (eg. managing organic matter, use of catchments), increasing the infiltration capacity and decreasing evaporation (eg. mulching) or draining excess water.

Drainage

Drainage system can aid the removal of surplus water. Drainage is one of the major inputs for increasing tea productivity. The necessity of disposal of surface and sub-surface water in the tea areas is a vital factor for growth of the tea plant. Removal of excess water from the soil prevents serious damage to the rooting system. Provision of efficient drainage system to maintain water table at a minimum depth of 100 cm is considered adequate for growing tea.

Uprooting

Top soil, being near the surface, is the major zone of root

development. It carries much of the nutrient available to plants and it supplies a large share of the water used by crops. Therefore, efforts should be made to minimize loss of rich top soil during uprooting of old tea bushes and at the time of clearing land for tea planting. It is also important to see that all roots are removed at the time of uprooting. Deep hoeing to a depth of 30 to 40 cm can help in removing left out roots of teas, shade trees and stones. Levelling of land of uprooted area before planting tea ensures that the new stand does not suffer localized waterlogged conditions.

Soil rehabilitation

It has been observed that the old tea land could have developed some toxic (allelopathic) substances since many years it remained under tea prior to uprooting. Further, lightening patches, root disease affected areas and compact soils will require rehabilitation. Guatemala (*Tripsecum laxum*), Norcot (Napier) (*Pennisetum purpureum*), Vetiver grass (*Vetiveria zizanioides*) and Mana grass (*Cympogon confertiflorus*) are excellent species for rehabilitation of soil. Planting of weeping love grass (*Eragrostis curvula*) has given good result at high elevation and slopes. Soil rehabilitation will help in improving the soil structure and also increase fertility of the soil. It also increases microbial activity of the soil. The plant species used for rehabilitation should have a deep and extensive fibrous root system and should yield large quantity of green material in a short period of time.

Mulching

This is one of the age-old proven cultural practices. Mulch can be defined as a shallow layer at the soil/air interface with properties that differ from the original soil surface layer. Mulching is an important technique for improving soil microclimate, enhancing soil type, structure and fertility, conserving soil moisture, reducing weed growth, preventing damage by impact from solar radiation and rainfall (erosion control), and reduces the respiratory rate of roots thereby affecting economy in the oxidation of carbohydrates. Mulching can ideally be done towards the end of the monsoon with Guatemala grass (*Tripsacum laxum*), weeping love grass (*Eragrostis curvula*), bracken fern, tea prunings, shade tree loppings and other plant materials. Effective mulching in young tea fields could be achieved with 15 to 25 tons of plant materials per hectare. It would be worthwhile planting Guatemala grass in large vacant patches to facilitate rehabilitation of soil and raising mulch material. Possible constraints to mulching are insufficient availability of mulch materials and pest problems (rodents, insects, fungi, persistent weeds).

References

1. Othieno, C. O. 1975. Surface run-off and soil erosion in fields of young tea. *Tropical Agric.* (Trinidad), **52**: 299-308.
2. Carr, M. K. V. and Stephens, W. 1992. Climate, weather and yield of tea. In: Tea: *Cultivation to Composition* (eds. K. C. Willson and M. N. Clifford). Chapman & Hall, London, pp. 87-35.
3. Arunachalam, K. 1995. *A Hand Book of Indian Tea.* All India Printing Press, Pondicherry, India. P. 271.
4. Owuor, P.O., Obaga, Samuel O. and Othieno, C. O. 1990. The effects of altitude on the chemical composition of black tea. *J. Sci. Food Agric.* **50**: 9-17.
5. Hipolito, M. C. 1999. Soil acidification: myths or reality. *ILEIA Newsletter*, **15** (1/2): 25-26.

Chapter 6

Cultivation and Maintenance

Selection of site

The area needs to be sufficiently isolated to ensure that there is no possibility of any pollutants or contaminations flowing or drifting into it from any known or unknown sources. There should be a buffer zone of sufficient width on all sides of the plot depending on the topography of the area, to ensure the above mentioned safety. The minimum width of buffer zone should be around 100 meters. A suitable buffer zone should also be maintained between organic and conventional units to prevent possible contamination of chemicals, if any. A detailed history of a period of about ten years, fully documented to give details of external inputs during pre-conversion period, should be maintained to facilitate inspection of organic tea cultivation. At the time of selection of site for establishment of new estates, the climatic requirements are the first consideration because these have great influence on growth, yield and quality of tea plant. If planting is done in the sloppy area then the extent of the slope should be assessed and the terraces, pattern of drainage etc should also be worked out.

Suitable sites for construction factory, workshop and housing should also be kept in mind at the time of site selection. Since organic tea estate should be a self-sustaining unit therefore the farm designing plays a vital role in optimizing the utilization of resources within the farm. The border trees, bunds, cattle shed, compost yard, storehouse, farm office etc must be suitably incorporated in the proposed organic tea plantation. The pattern of roadways should be designed to ensure easy accessibility to all parts of the land, transportation of green leaf, processed tea, plants, manure and personnel throughout the year.

Preparation of land

Virgin forest land is usually the most suitable for organic tea cultivation. Grassland and areas which have grown other crops can also be used. The area intended for planting is first cleared of all kind of growth. In general, burning of forest growth is not advocated as it makes the soil alkaline. Trees if any should be killed in advance and positively ring-barked or frilled. It minimises the risk of root disease. Stones, if any, should also be removed. Pits created while uprooting the roots should be filled up with soil. Trees and other unwanted materials should be moved off the field. These operations should preferably be carried out in the dry season. When land is cleared for planting, leguminous cover crops may be sown before rains. While selecting species of the cover crops, the following characteristics are to be considered.

- It should have the characteristics of yielding large quantity of green material in a short span of time and ability in suppressing the weeds.
- The plant should have more leafy growth than woody growth which enables rapid decomposition.
- The plant should possess deep and fibrous root system.

The land preparation should be such that at the end of rehabilitation period, the fertility level of the soil is optimized and no residues of any chemical from previous treatments are left.

Soil conservation measures

Adoption of proper soil and water conservation measures is essential for organic tea cultivation. After clearing the land, soil conservation works should be the major concern and must be put in hand as early as possible. The top soil is valuable for growth of plants and conservation is therefore necessary. Soil should be adequately protected from erosion. Many tea estates in Darjeeling hills in North East India, the Nilgiris and the high range in South India have recurrently suffered from soil erosion problem. Soil erosion control in tea lands in Sri Lanka, India, Southern Africa and Kenya has been discussed in the past.[1-4] The important soil conservation measures consist of terraces, contour wall, micro-catchments etc and should be carried out wherever necessary before the tea is planted. Other control measures, such as proper drainage at the upstream and towards safe outlets, followed by good vegetative growth on exposed surfaces, planting of deep-rooted local grasses, shrubs and trees which provides a binding force to soils exposed to the atmosphere, double hedge or staggered system of planting, practice of minimum tillage should be taken up immediately. Mulching is also protecting the soil from sunshine and impact of raindrop.

Establishment of vegetative barrier is an important component of soil conservation measures. This is less expensive. Weeping love grass (*Eragrostis curvula*) has been found to give satisfactory results when planted along embankments. In this context vetiver grass technology assumes significance. This is the establishment of a narrow (less than one metre wide) live grass barrier, in the form of a hedge, across the slope of the land. This is effective on slopes from less than 1 to over 100 percent. A well-established grass hedge will slow down run off of rainfall, spreading it out evenly and will trap runoff sediments to create natural terraces. *Vetiveria zizanioides*, when planted will quickly form a dense, permanent hedge, which will reduce the velocity of water flow, divert run out water and form an effective filter. It does not compete with tea in the field. Its sharp leaves and aromatic roots repel rodents. Once vetiver is established in the field, it is generally unpalatable to livestocks. However, the young leaves are palatable. It will grow in a wide range of climates and is known to grow in areas with average annual rainfall between 300 and 6000 millimetres and with temperature ranging from -15 ^{0}C to 55 ^{0}C [5].

Terraces

The construction of terraces in sloping land is essential for minimization of rapid loss of surface soil. Terraces should follow the contour of the slope and the upper surface of the terrace may slope slightly towards the hillside to prevent soil wash. Growth of grass may be allowed along the edges and side of the terraces.

Contour planting is helpful to a terrain that cannot be mechanically terraced. Contour planting is done first marking a master row between a pair of graded contour drains and then tea rows are planted parallel to the master row. The master row is marked as (i) if a pair of graded contour drains converge in the direction of flow, the master row is made parallel to the top or higher contour drain, and (ii) if a pair of contour drains diverge in the direction of flow it is made parallel to the bottom or lower contour drain. A separate master row is marked between each pair of contour drains.

Cover crops

When land is cleared and after earthworks, leguminous cover crops such as *Mimosa invisa* Mart., *Stylosanthes guyanensis* SW var. *gracilis* (HBK) Vog., *Calapogonium mucunoides* Desv., *Centrosema pubescens* Benth., *Pueraria phaseoloides* (Roxb.) Benth., may be sown before rains. *Mimosa invisa* and *Desmodium* sp. in combination with Guatemala (*Tripsacum laxum* Nash) would be ideal. The growth of cover crop can act as a barrier to protect the soil of the newly-opened land against erosion. It also suppress weed growth, improving soil structure and fertility, and helpful in keeping down the soil temperature during summer months. Cover crop can be grown in-between the tea rows and in walking rows. Seeds can be sown 45 cm from tea rows. Plants can be pulled out when seeds become mature. Non-leguminous cover crops have also been used with success. *Polygonum*, *Oxalis* is of considerable value in south India, Sri Lanka and East Africa. In Kenya, Oats (*Avena sativa* L) and buckwheat (*Fagopyrum esculentum* Moench.) are widely used as cover crop for higher and lower altitudes respectively. These crops can be established easily by direct seeding and dieback after one year. *Crotalaria* is also widely used as nurse crop planted between rows of young tea plants.

Drains

High water table resulted in frequent waterlogging in the Assam and bheel soils of Cachar tea areas of North East India, and certain swampy or river beds in a number of tea-growing countries. In Darjeeling, both surface waterlogging and high water table due to lateral spring are observed. The surface waterlogging is the result of localised depression formed near the bushes. The water table rises to undesirably shallow levels and affects normal growth of tea plants due to reduced rate of mineralisation, lower microbial activity, and high denitrification with lack of aeration in the soil, low soil temperature under wet conditions, and poor response to applied compost. Waterlogged plants exhibit certain symptoms such as scanty shoot production, yellowing of leaves and premature defoliation, dieback in plucking points and poor recovery after pruning, infection by violet root rot and red rust diseases, loss of vigour etc. To cope with this situation it has become necessary to provide artificial drainage facilities to drain excess water below the root zone. Drains should be dug in winter when soil moisture is low. Digging should be started from outfall and go upward. Besides boundary drains, contour drains must be dug at regular intervals. Drains are to be traced along the contour of the land in a gradient of 1:120. The length of the drains, and the distance between them depend on the lie of the land. In a moderately undulating ground it could be spaced at about 12.2 m (40 feet), in a mild slope at 18.3 m (60 feet) and in steep terrains at 6.1 m (20 feet). It would be beter if these drains provided with lock and spillover arrangements to arrest the velocity of the water and connected to leader drains are taken wherever there is natural converging of slope. The vertical drains should be taken with inward sloping as otherwise the velocity of water flowing through the drain will increase commensurately in proportion to the length of the drain.

Planting

This operation basically determines the growth and yield level of a plant. It is always desirable that before planting, soil testing is done to ascertain the fertility status of the soil and for the presence of any chemical residues in the soil. The site for planting can be marked by a stick. After planting, the field should be mulched (6 to 8 cm thick) in absence of cover crop. But it must not touch the stem of the young plants. The organic techniques during and after planting are to be critically observed ensuring that no external chemicals are applied to make the plants grow faster or to make them stress-free following their planting.

Planting distances

It is difficult to generalise on the number of plants ideal for a particular field area to all situations because of the involvement of number of diverse variables. Planting density is however an important determinant of yield. For high yield, complete ground cover is important. In recent years, the trend allover is to go for closer spacing in order to achieve high yield. Generally, close spacing yield a high crop in early years and expedites recovery of the heavy cost of planting operations notwithstanding better conservation of soil and reducing the cost of weed control. The main disadvantages are suppression of overall growth of the plant and restricted development of root system. There are some other factors also which have a bearing on spacing. For example, spreading habit of clone (compact or spreading frames), steepness of the slope, fertility status of the soil, machine harvesting etc can dictate the optimum spacing. Nowadays, single hedge and double hedge system of planting are very common. The number of plants per hectare that can be accommodated at different spacing are furnished in Table 6.1. The double hedge planting is recommended for flat or gentle sloppy area and it should be avoided in steep land where plucking would be difficult. The rows are laid in east west direction for maximal utilization of the solar energy.

Planting hole

It should be of adequate size. Planting hole of 45 cm wide and 75 cm depth are better for establishment of young plants. When the holes are being dug the top soil should be placed on one side and subsoil on the other side. In clayey soil and drought-prone areas, trench planting in place of holing will be advantageous. Trenches of 30 cm wide and 45 cm depth along rows are dug which provide lower

Table 6.1. Plant population per hectare at different spacings.

Type of planting	Spacing (cm)	Plants
Square	120 x 120	6274
	135 x 135	5315
	150 x 150	4305
Triangular	120 x 120	7770
	135 x 135	6139
	1450 x 150	4353
Single hedge	90 x 60	18518
	110 x 60	15151
	120 x 60	13888
	120 x 75	11111
	120 x 90	9259
Double hedge regular	110 x 60 x 60	19607
	120 x 60 x 60	18518
	120 x 75 x 75	13675
	120 x 90 x 90	12345
Double hedge stagger	110 x 60 x 60	20575
	120 x 60 x 60	19379
	120 x 75 x 75	14414
	120 x 90 x 90	13888

volume of soil for expansion of roots. Filling the holes should be done with top soil as far as possible. Sufficient surface soil has to be collected for filling the holes after removing stones and roots. A tall strong marking stick should be put in at the middle of the filled pit to indicate its position. Vermicompost or the planting mixture per pit is made up of well rotten dry cattle manure (4-5 kg) plus rock phosphate (30 g) may be incorporated into the planting pits.

Planting time

In general, a cool, humid climate and moist soil are favourable for high success after planting. Planting is commenced in Nilgiris, Anamallais in South India and Darjeeling with the onset of pre-monsoon showers during the month of May. In the eastern part of Assam, winters provide sufficient moisture in the soil and cool misty weather accompanied by occasional showers makes planting possible during late autumn and early winter. The middle and western part of Assam including Cachar and tea areas in the plains of West Bengal experience droughty conditions from November to April and thereby planting is safe during May - June. In most of the tea areas of Indonesia and Sri Lanka, planting can be done at any time of the year as moderate temperature, adequate and well distributed rainfall prevails most of the year. In Japan, late March to early April is the suitable time for planting. The tea areas of East Africa experienced deficit of water in the soil, low temperature, and high evapotranspiration loss due to strong wind during the early part of the year. In Central Africa, suitable period for planting is even more limited.

Sleeved plants planting

Vigorous plants of uniform height can be taken out from the nursery when sleeves are wet. There should be profuse feeder roots nearer the bottom-end of the sleeves at the time of planting. The main stem at the collar region should be about pencil thick and brown; it should be straight and errect. Weak and smaller plant may create problems at the time of weed control, training and susceptible to pests. Sleeved plants should preferably be carried in boxes to the planting site to minimise disturbance of the soil which could damage the roots. A vertical cut is made at the bottom half of the sleeve and the space around the sleeve are gradually filled with the cylinder of soil without being broken at the time of this operation. When the hole is thus partially filled the bag is slit along its full length and carefully removed. The gap around the upper part of the soil core is also filled with soil without causing any damage to the soil core. If the roots overgrow the size of the sleeve or *bhetis*, the roots at the bottom of sleeve or *bheti* should be cut to avoid twisting. Depression if any at the collar due to poor ramming may creates localized water-logging resulting in mortality. Further, ramming the soil too hard causes compaction which is detrimental to growth.

Stump planting

A stump with plenty of food reserve gives a vigorous shoot with healthy leaves. Stumps that are at least 1 cm in diameter at the collar and have been unshaded for 5 - 6 months to build up starch reserve are ready for planting. Stumps after pulling out from the nursery should be cut off 10 cm above the soil level, and the root 45 cm below soil level. Thin lateral roots should also be removed. Seedling make single tap root. Special attention should be given to ensure that the tip of the tap root is in actual contact with the soil at the bottom. Drying out of the stump should not be allowed. It is always advisable to plant the stumps soon after pulling out from nursery beds.

Infilling

Tea bushes in the established field may die for a variety of reasons such as poor drainage, pests and diseases, debility, lightning etc. Infilling is the replacement of dead plants by new one and it should be persisted with until the stand of young tea is completely filled. Vacancies encourage growth of weeds in the gaps and its presence in significant number reduces overall production. Infilling of vacancies is better done in the first two years when the young plants have not spread much so that the infill will have more space to grow freely and therefore a better chance of survival. After planting of infill, the ground should be mulched with any suitable vegetative material in order to reduce the loss of soil moisture and add organic matter to the soil. Green mulch has to be kept 10-15 cm away from the collar of infill/interplant to avoid heat injury and termite attack.

Infilling of individual vacancies or in blocks should preferably be taken up in the year of heavy, medium or light pruning. Strong plants with good root system of appropriate clones or proven seed stock should be used for infilling in order to compete against the established plants around. While infilling, a large hole should be dug to ensure presence of no live roots of established tea close to the new plant. Preferably two plants per vacancy in the large pit where hard pan has been broken and adequate cattle manure given should be planted. The side branches of the neighbouring bushes if encroaching into the vacant space to be occupied by the infill can be trimmed and infill tipped a little higher in the first two years or so. Infill should be planted at a closer spacing than the established plants. Pegs should be put around the infill which would protect plants from trampling. Proper attention on adequate post planting care is required for successful establishment of infill.

Shading

The young plants require protection from the bright sun and dry wind. Shading trees have a great importance to the organic cultivation of tea. The following is a list of their positive effects:

- Regulates temperature and increases humidity at bush level
- Minimises loss of water through evaporation and transpiration
- Serves as wind break

- Reduces ultra-violet radiation injury and incidence of mites
- Its litter and loppings contribute considerably to the soil organic matter pool and is useful in maintaining the soil fertility and recirculation of nutrients
- Generates additional income by way of timber and fuel
- Minimises soil erosion and maintains the ecological equilibrium
- Helps in getting better distribution of crop
- Encourage beneficial insects to settle
- Create a pleasant atmosphere for the pluckers

While selecting species of shade trees, the factors like growth habit, canopy architecture, depth of root system, economic life and commercial timber value, less pest and disease incidence, tolerance to wind and frost, suitability to the soil and climate of the locality etc. should be considered. Most of the species that are used as shade are leguminous. Leguminous trees can fix significant amount of nitrogen from air and provide for the tea. In North East India the permanent shade tree species found suitable are: *Albizzia chinensis* (Osbeck) Merr., *A. lebbek*, *A. odoratissima* (L.f.) Benth., *Acacia lenticularis*, *Derris robusta* and *Dalbergia sericea* (Table 6.2). In South India, *Grevillea robusta* (Silver oak), a non-leguminous plant is best suitable shade tree species. *Acacia mearnsii* de Wild. (Black wattle) and *A. decurrens* (J. Wendl.) Willd. can be used in frost-prone areas of high elevation planting districts of Nilgiris and High Range. The shade trees should be carefully regulated by pollarding the trees, where possible, and thinning out the branches as required. The young shade trees planted are usually taken a few years to provide effective shade to the newly-planted tea. It has been observed in different places that tea was planted between the lines of green crops like *Crotalaria anagyroides*

Table 6.2. Proliferation of different varieties of shade trees in various tea-producing regions.

North East India	South India/ Sri Lanka	Indonesia	East Africa
Albizzia chinensis *Albizzia odoratissima* *Albizzia lebbek* *Acacia lenticularis* *Dalbergia sericea* *Derris robusta* *Erythrina indica*	*Erythrina ssp.* *Gliricidia ssp.* *Grevillea robusta*	*Albizzia chinensis* *Albizzia moluccana* *Albizzia falcate* *Erythrina ssp.* *Leucaena glauca*	*Grevillea robusta* *Albizzia gummifera* *Albizzia adiantifolia* *Gliricidia maculate*

HBK., *Priotropis cystisoides* (Roxb.), *Desmodium gyroides* (Roxb. ex DC.), *Sesbania* sp., *Trifolium ripens* L. etc for protecting the young tea plants from direct sun rays.

If the leaf temperature goes above 30^{0}C then shade becomes necessary irrespective of location. Therefore, shade is not necessary in Kenya, Uganda, Tanzania, high and mid-elevation tea areas of Sri Lanka, high elevation tea estates of Darjeeling, South India and Indonesia where ambient temperature does not often exceed 30^{0}C. But Tea Research Institute, UPASI recommends regulated shade in Nilgiris and under all situations not because of leaf temperature but in consideration of other factors like severe intensity of sun rays and radiation damage and leaf chlorosis were noticed in the tea plantations situated above 5,000 ft. During summer in South India, draughty conditions are normally experienced between December and March, at times extending over up to end of April or early May. Besides, the soil moisture stress, the intensity of drought is aggravated by the low levels of atmospheric humidity during February and March and high diurnal temperature that surpasses even 35^{0}C in the mid-elevation estates during summer months. The need for shade in tea in South India is thus well recognized.

Despite planting green crops, another practice which is popular nowadays in India is the planting short-duration,quick growing plants to provide shade during the early years.

Indigofera teysmanii, Tephrosia candida (Roxb.) DC. (Bogamedeloa) and *T. vogelii* Hook. f. (African medeloa) is extensively used as temporary shade tree species both in North East and South India. *Leucaena leucocephala* (Lam.) de Wit (Su-babul) are also used in a very limited scale as a temporary shade. *Sesbania cinerascens* Welw. ex Bak., *Crotalaria agathiflora, Acacia elata* A. Cunn. ex Benth. can be planted at a higher altitude in South India. These are initially planted at close spacing and keeping them trimmed and lopped from time to time and finally removed when permanent shade species take over.

Spacing of shade

- *Temporary shade*: Initially 2 m x 2 m and later on thinned out to 4.5 m x 4.5 m.
- *Permanent shade*:

Light shade: 15.24 m x 15.24 m (triangular) or 13.7 m x 13.7 m (Square or wider).

Medium shade: 13.7 m x 13.7 m (triangular) or 12.2 m x 12.2 m (square).

Heavy shade: 10.67 m x 10.67 m (triangular) or 9.14 m x 9.14 m (square or closer).

Planting of shade trees

The planting pit should be 90 cm deep and 60 cm wide. It would be ideal if the bottom soil of the pit is loosened to a further depth of 15 cm. A manure mixture composed of 10 to 15 kg dry well rotten cattle manure (or 1.0 - 1.5 kg of oil cake), 1 kg rock phosphate, 2.5 kg of wood ash and 1 kg slaked lime to be added in pits at least seven days prior to planting. Sleeve grown seedling or with a *bheti* (60 cm long and 30 cm diameter) is the best method of planting. The roots found at the bottom and outside in sleeve-raised and *bheti* plants respectively should be cut off. The shade plant at the time of transplantation in mature tea areas should be 1.8 m to 2 m tall and 60 cm to 90 cm in young tea field. Woven bamboo guards help to protect young shade plants from cattle and other physical damage.

Irrigation

Tea is usually a rain fed crop. Tea-growing areas of few countries like Sri Lanka, Indonesia and Kenya are getting some rains throughout the year but the major regions where irrigation is necessitated by a prolonged dry season are Bangladesh, India, Malawi, Tanzania, and South Africa. Irrigation is also practiced for short periods in parts of China, Japan, and some other tea-growing areas. A perennial source of water free from pollutants for irrigation and large-scale compost preparation are essential for organic tea cultivation. The young plants particularly after planting in the field should be protected by a suitable shading material and sub-soil irrigation may be given wherever possible. However, this is an expensive measure and can be practiced only if returns are attractive. Drip irrigation, though it is expensive, is being successfully adopted in many gardens, which can achieve high yield and good price for their tea.

Training of young tea

The early management of young tea is aimed at modifying and training it to form a low bush of good spread and of appropriate height for plucking. Therefore, it is most important to establish a well-developed primary frame and branch system to ensure a complete ground cover as early as possible. These objectives can be achieved either by decentering and pruning, or by bending or pegging or by a combination of decentering and bending [6]. Recommended procedures for different regions vary. Variations may be useful keeping the rates and habits of growth of particular clones. The leader stem of the young plants is cut at a height between 7 and 20 cm from the ground (4-6 months after planting in South India). Later, these plants are tipped at 35 cm height and second tipping will be at 50 cm in South India. The young plants shoud be manured and mulched. Necessary protection against drought and pests are important.

Pruning

As the age of tea bushes from pruning advances, size and weight of harvestable shoots declines due to the reduction in the vascular supply to growing buds. An extreme case of growth retardation is evidenced in the higher proportion of *banji* shoots at the plucking table. Also, the upper part of the plants below the plucking table becomes congested with stems and knots where many new stems have originated. Despite the 'maintenance foliage', ages and losses efficiency, lower leaves dropping off, lead to uneconomic harvesting. Pruning at this stage becomes necessary by which these irregularities could be remedied. Pruning is an important practice in tea cultivation which involves removal of certain amount of growth of the bush.

Objectives of pruning

- Maintaining convenient height to facilitate harvesting operation.
- Checking reproductive growth and stimulating vigorous shoot system, rationalising wood / shoot ratio.
- Removing dead, diseased and defunct wood thus help rejuvenate bushes that have crossed the period of maximum productivity.
- Minimising formation of *banjhi* (dormant shoots) and spreading of pests and diseases.
- Renewing the growing branches and maintaining the sufficient volume of maintenance foliage to meet the physiological needs of the plant.
- Controlling crop during rush periods and to achieve better crop distribution.
- Maintaining quality in processed tea.

The interval of time between two successive prunes is called a pruning cycle. The length of pruning cycle may vary from 1 to 6 years depending on growth pattern of the tea cultivar, soil-climatological conditions, plant variety, height of pruning, quality and quantity of crop needed and availability of plucker. Further, most of these factors interact to influence productivity, therefore, without considering of these interacting processes it is very difficult to generalise anything on an appropriate pruning cycle. The details of pruning cycle and types of pruning followed in different tea-producing countries have been discussed by Ghosh Hajra [6]. No fundamentally different pruning measures are used in organic tea estates to those carried out in conventional tea plantations.

Diversification strategies

A diversification of the agro eco-system is one of the basic requirements in organic agriculture. It is not unusual especially for small growers that income be entirely dependent on the sale of harvest from one single crop. When prices then fall, this will inevitably lead to very serious problems. Diversified agro eco-systems consisting of a variety of crops that can either be sold (local market, export) or are for own-use (food, fodder, building material and fuel) have a far better crisis-resistant structure. Yet such diversified systems require an efficient planning and organisation of the available workforce. A new crop is to be introduced into the plantation. Before introduction, types of other crops can be integrated into the tea plantations for each individual production site needs to be decided. Further, technical requirements for post-harvest treatment (e.g., drying, storage etc.) and availability of manpower due to parallel cultivation of different crops must be ascertained.

The crops which can provide food for workers' families, to be sold on the local market or used as additional cash crops should be preferred at the time of selection of crop. The cultivation of spice plants, viz. cardamom and ginger in Darjeeling or nutmeg nuts and pepper in Sri Lanka are worth mentioning. Vanilla can also be easily integrated into organic tea plantation (climb up the shade trees).

References

1. Krishnarajah, P. 1985. Soil erosion control measures for tea land in Sri Lanka. *Sri Lanka J. Tea Sci.* **54**(2): 91-100.
2. Barua, D. N. 1989. *Science and Practice in Tea Culture.* Tea Research Association, Calcutta-Jorhat, India. p. 493.
3. Grice, W. J. 1983. Soil and water conservation at planting. *Tea Research*

Foundation (Central Africa) Quarterly Newsletter **69**: 13.

4. Othieno, C. O. 1992. Soil. In: Tea*: Cultivation to Composition* (eds. K. C. Willson and M. N. Clifford). Chapman & Hall, London. pp. 137-172.
5. Anonymous, 2000. *Vetiver Grass-The Hedge Against Erosion*. The World Bank, Washington, D. C. p. 78.
6. Ghosh Hajra, N. 2001. *Tea Cultivation: Comprehensive Treatise*. International Book Distributing Co., Lucknow, India. p. 518.

Chapter 7

Nutrient Management

Organic farming is often understood as a form of agriculture with use of only organic inputs for the supply of nutrients and management of pests. In fact, it is a specialized form of diversified agriculture, wherein problems of farming are managed using local resources alone. The term organic does not explicitly mean the type of inputs used; rather it refers to the concept of farm as an organism. Often organic agriculture has been criticized on the grounds that with organic inputs alone, farm productivity and profitability might not be improved because the availability of organic sources is highly restricted. True organic resources availability is limited; but under conditions of soil constraints and climate vagaries, use of organic inputs proved more profitable compared to agrochemicals [1].

The perennial tea plant requires a considerable amount of nutrients for growth. High quantities of nutrients are lost through the continual plucking of tea leaves. Manuring aims at making available all the essential nutrients deficient in the soil at optimum quantities and to return the nutrients removed by using the manures of organic origin demanded by the tea bushes for a sustainable productivity level. Table 7.1 and 7.2 provides average nutrient losses on various tea cultivation regions, which are based on studies carried out on conventional tea plantation (therefore, the values can only be approximately used for the situations on organic tea plantations).

In the organic tea estates, nutrients are being supplemented using rock phosphate, well composted farmyard manures, biogas slurry, neem cakes, oil cakes, lopping of green manures, shade tree droppings etc. In general, most of the workers in the organic tea estate are supplied with a cow or a goat per household. This provides milk for the family and any surplus is sold. Organic tea growers in Sri Lanka are now encouraged to use more goat dung as it gives the plants enough copper and zinc (http://forests.org). The cow dung along with kitchen waste, after composting, forms a good source of nitrogen. The plantation buys the compost or the organic manure from the workers. The workers are encouraged to grow fodder for their cows on areas that are unsuitable for tea.

Growing leguminous crops (shade trees, cover crops, green crops etc) would also be a good source for the supply of nutrients. Two-tier mixed shade canopy as source of vital plant nutrients is also important for organic tea estates. Shade trees contribute substantial amount of leaf litter (ca. 15 tons/ha) every year, which contributes to the humus content of the soil and recycling of nutrients from within the deeper layers of the soil. Well decomposed organic matter and farmyard manure, applied at the right soil moisture condition would not only improve soil texture and structure but also provide necessary resistance to the crop against pest and diseases. Bone meal, fish meal and other manures of organic origin can also be used if available from unpolluted environment; wood ash can also be applied. In Tanzania, locally known 'sunflower cake' is certified as organic and is made up of organic chicken manure imported from South Africa, combined with fish meal from Lake Victoria's fisheries (www.clipper-teas.com). Tea waste may be used as such or employed as a base for preparing bulky organic manure by mixing thoroughly and uniformly with straw, green manure, fallen leaves and light prunings. Well composted tea waste is odourless, innocuous and of high manural value. The estimated nutritional values of some organic waste materials are presented in Table 7.3. There are some restriction in using some of these inputs for use in organic production. In the Table 7.4 'Restricted' means at the conditions and the procedure for use shall be set by the accredited certification programme. Factors such as contamination, risk of nutritional imbalances and depletion of natural resources shall be taken into consideration.

It is, therefore, vital for the concerned authority of the tea estates to establish a fertilization programme right at the beginning of the conversion that places core emphasis on the production and the subsequent usage of organic substances, as the most important source of nutrients. Furthermore, the nutrient content of the soil should be analysed regularly, particularly the supply of potassium, phosphorous and magnesium (also for trace elements). If deficiencies arise, additional fertilisers approved for use on organic plantations (e.g. rock phosphate, potassium sulphate, kainit and sylvit) could be applied. In order to maintain an ideal pH value, liming (e.g. with dolomite meal) may be necessary. In the case of extremely low pH values (risk of Al toxic ity), the use of gypsum ($CaSO_4$) is also permitted. However, it is advisable to get prior approval from the certifying agency before application of these fertilizers.

Table 7.1. Average nutrient losses on various tea cultivation regions (extraction in kg for 1000 kg/tea/ha/year.

Region	Nitrogen (N)	Phosphate (P_2O_5)	Potassium (K_2O)
North India	50	10	20
South India	65	15	35
Sri Lanka	45	8	21
East Africa	42	6 - 8	24

Source: Swiss Import Promotion Programme (www.sippo.ch)

Table 7.2. Nutrient loss throughout a pruning cycle of 3 years.

Loss (kg/ha)	Nitrogen (N)	Phosphate (P_2O_5)	Potassium (K_2O)
Pruning	785	135	570

Source: Swiss Import Promotion Programme (www.sippo.ch)

Essential plant nutrients

Green plants are autotrophic since they manufacture their own food from CO_2 and H_2O in presence of sunlight. But for inorganic matter they depend on the outside world. Plants obtain a wide variety of inorganic elements from the soil. These inorganic elements which are usually grouped together under the class "mineral elements or mineral nutrients" and the nutrition of mineral nutrients is called mineral nutrition. The source of these elements is mainly the ions present in the soil solution and absorbed by the root system.

Among the essential elements, oxygen, hydrogen, carbon are obtainable from air and water and need no specific consideration. The other thirteen mineral nutrients are essential for the growth and development of all plants including tea. For reasons of convenience and tradition, these are usually categorised as follows.

Macro		Micro	Beneficial elements
Primary	Secondary		
Nitrogen (N) Phosphorus (P) Potash (K)	Sulphur (S) Calcium (Ca) Magnesium (Mg)	Zinc (Zn), Iron (Fe), Manganese (Mn), Copper (Cu), Boron (B), Molybdenum (Mo), Chlorine (Cl)	Aluminium (Al), Silicon (Si)

Table 7.3. Nutrient status of some organic waste materials.

Category	Source	Nutrient content (%)		
		Nitrogen (N)	Phosphoric acid (P_2O_5)	Potash (K_2O)
Animal waste	Cattle dung	0.3 – 0.4	0.10 – 0.15	0.15 – 0.20
	Cattle urine	0.80	0.01- 0.02	0.5 – 0.70
	Sheep and goat dung (mixed)	0.65	5.5	0.03
	Night soil	1.2 – 1.5	0.8	0.5
	Human urine	1.0 – 1.2	0.1 – 0.2	0.2 – 0.3
	Leather waste	7.0	0.1	0.2
Compost	Farmyard manure	0.5 – 1.0	0.15 – 0.20	0.5 – 0.6
	Poultry manure	2.87	2.90	2.35
	Town compost	1.5 – 2.0	1.0	1.5
	Rural compost	0.5 – 1.0	0.2	0.5
	Water hyacinth compost	2.0	1.0	2.3
	Tea waste compost**	3.60	0.60	2.30
Oil cakes	Castor	5.5 – 5.8	1.8	1.0
	Coconut	3.0 – 3.2	1.8	1.7
	Cotton seed	3.9 (6.5)*	1.8 (2.8)*	1.6(2.1)*
	Ground nut	4.5 (7.8)*	1.7 (1.7)*	1.5 (1.4)*
	Karanj (*Pongamia Pinnata*)	3.9 – 4.0	0.9 – 1.0	1.3
	Neem	5.2	1.0	1.4
	Niger	4.8	1.8	1.3
	Mahua butter tree (*Bassia latifolia*)	2.5 – 2.6	0.8	1.8
	Rapeseed	5.1	1.8	1.0
	Linseed	5.5	1.4	1.2
	Safflower	4.8 (7.8)*	1.4 (2.2)*	1.2(2.0)*
	Sesame	6.2	2.0	1.2
Animal meals	Blood	10 – 12	1.2	1.0
	Meat	10.5	2.5	0.5
	Horn and hoof	13.0	0.3 – 1.5	-
	Raw bone	3 – 4	20 – 25	-
	Steamed bone	1 – 2	25 – 30	-
	Fish	4 – 10	3 – 9	1.8
Miscellaneous Materials	Paddy husk**	0.15	0.13	0.22
	Paddy straw**	0.50	0.07	0.60
	Maize straw***	0.40	1.50	1.60
	Thatch grass**	0.50	0.10	0.50
	Guatemala grass***	1.6	0.60	1.70
	Napier grass***	0.8	0.40	1.30
	Mimosa***	2.6	0.60	1.00
	Tea pruning litter**	2.0	0.50	1.50
	Shade tree droppings**	2.5	0.70	0.80
Wood ash	Ash, coal	0.73	0.45	0.53
	Ash, household	0.5 – 1.9	1.6 – 4.2	2.3 – 12.0

Figures in ( ) are for decorticated material,

Source: Ref. 2, 3**and 4***.

The role of Al and Si in plant metabolism is not clear. However, silicon is well known for its oxydising power making native phosphate easily available. Aluminium has also been found to be an essential nutrient for tea [5]. All these nutrient elements carry out specific functions in the plant. Though they are equally essential they are required and absorbed by the plant in widely different amounts. For example, 100 kg processed tea may contain 4 kg N but only 4 grams of Zn (1000 fold difference) and yet if the tea bushes do not get these 4 grams of Zn, they neither make efficient use of other nutrients nor produce high yields [6].

Effect of individual nutrients

Nitrogen

The supply of nitrogen in organic farming is usually provided by legumes. Through a symbiosis with nodule bacteria, these plants are capable of fixing atmospheric nitrogen and making it available to plants. There are other bacteria that can also fix nitrogen (Actinomycetes which are present in dead wood, soil-bound bacteria viz. *Azotobacter* or *Beijerinckia*, which live in association with the tropical fodder-grass *Paspalum notatum* and other gramineae).

Phosphorous

The phosphate (P) content of the soil vary just as much as the availability of phosphates for the plants (e.g. tropical soil with its high acid, iron and aluminium content has a very high rate of P fixation, thus applied P-fertilisers become unavailable for the plants). On organic farms emphasis is placed upon increasing the availability of the phosphate content of the soil for the plants. This is achieved by a biological conversion of unsoluble to soluble P-compounds in the soil (enzymes and plant acids):

- In fertile soils with sufficient organic matter the conversion to soluble P-compounds is most efficient. Fertile soils have a more intensive growth of roots and subsequently a more developed network of fine roots increasing the interaction between fine roots and phosphate compounds in the soil. Acid released by the fine roots than dissolve the fixed P-compounds and improves the availability for plants.
- Fertile soils with high organic substance content encourage the growth of VA-Mycorhizza, a fungus that lives in a symbiosis

with plants having a high capacity to dissolve fixed P-compounds.

- Organic matter (mulching material, compost) increases the availability of phosphates
- High pH values and poor phosphate availability can be alleviated by applying silicates.

Adding rock phosphates is still allowed on organic farms. In the new clearings, phosphorus may be applied@ 90 kg/ha/year (23 kg of rock phosphate or 35 g of mussorie phosphate/plant). It should be applied by placement at 10 to 22.5 cm depth in holes made by crowbar on both sides of tea plants on the upper side of the slope. In the established tea fields, soil available P should be maintained at or above 22 ppm. Phosphorus may be applied @ 60 to 80 kg ha^{-1} in the first and third years as placement based on the value of soil analysis.

Potassium

Requirement of potassium (K) in organic tea is usually met by the addition of wood ash from untreated wood. It should preferably be broadcasted during dry period 15 cm away from the collar. It should only be applied in mature tea fields except in pruned year. If potash deficiency is continued to be noticed and the exchangeable potassium level in the soil is found low by soil analysis, patentkali, a naturally mined potash fertilizer (30 percent K_2O and 10 percent MgO) need to be applied @ 200 kg ha^{-1}. Certain potassium salts (Muriate of potash/ potassium chloride is not allowed) with low chlorine content is also permitted to use. The following strategies to ensure a sufficient supply of K for the plants are important in organic farming:

- Regular applications of organic matter will improve the absorption of potassium in the upper soil layers, where it can be reached by the plants' roots.
- Use of deep-rooting plants to mobilize K in lower soil layers.
- A permanent mulching layer in order to reduce leaching of K especially in the heavy precipitation area (http://www.naturland.de).

Symptoms of nutrient deficiency

When plants are grown under inadequate supply of any one or more of the essential nutrients can in no case give satisfactory yield

Table 7.4. Fertilizers and soil conditioners.

Substances description, compositional requirements	Conditions for use
Matter produced on an organic farm unit	
* Crop residues, green manure, Straw and other mulches	
* Slurry, urine, farmyard manure	
* Vermicast and vermicompost	
Matter produced outside the organic farm unit	
* Slurry, urine, farmyard manure, straw, vermicasts	Restricted
* Compost made from any carbon-based residues	Restricted
* Blood meal, meat meal, bone meal, feather meal without Preservatives	Restricted
* Fish and fish products without preservatives	Restricted
* Guano	Restricted
* Human excrement	Banned
* By-products from the food and textile industries of biodegradable material of microbial, plant or animal origin without any synthetic additives	Restricted
* Peat without synthetic additives (prohibited in soil conditioning)	Restricted
* Saw dust, wood shavings (wood comes from untreated wood)	Restricted
* Seaweed and products of seaweed obtained by physical processes, extraction with water or aqueous acid and/or alkaline solution	Restricted
* Sewage sludge and urban compost	Restricted
Minerals	
* Basic slag	Restricted
* Calcareous and magnesium amendments	Restricted
* Mineral potassium with low chlorine content (e.g. sulphate of potash, kainit, sylvanite, patenkali)	Restricted
* Natural phosphate, pulverised rock, trace elements, wood ash from untreated wood	Restricted
* Calcified seaweed, calcium chloride, limestone, gypsum and chalk	
* Magnesium rock, sodium chloride	
Other products	
* Biofertilizers	
* Biodynamic preparations	
* Plant preparation and plant extracts	

unless the deficiency is corrected. The deficiency symptoms of various nutrients are described below.

Nitrogen

The yellowing of leaves (chlorosis) and finally pale are the symptoms which is usually prominent in lower leaves of current season's growth. Leaves become slender, small and fall off. The growth of plant slows down, internodes shorten and production of crop decline. Sometimes the leaves, petioles, veins and stems of the nitrogen deficient plants show purplish pigmentation due to synthesis of anthocyanin.

Phosphorus

Deficient leaves become dull in appearance and lose the glossiness. Sometimes leaves are mottled with yellow spots showing bluish green with purplish tints. Premature leaf fall and stunted growth are also noticed.

Potash

Mottled chlorosis followed by necrotic shrivelling of the margins of the leaves, premature leaf fall and development of twiggy wood are the symptoms of deficiency. The bark of the stem becomes silvery white. Sometimes there will be profuse branching of shoots at the top and little or no branching from the lower older stems. The recovery from pruning is slow and irregular.

Magnesium

Yellowing of leaves near mid-rib and spreads gradually between the main lateral veins. The older leaves are affected earlier and fall off easily than the younger leaves.

Sulphur

Deficiency of sulphur results in chlorosis. The yellowing first appear in young leaves. In some cases most of the leaves show chlorosis at the same time. The internodal distance is reduced, leaves scorch and fall off and stems die back from the tips. The edge of the leaves curl up and tip turn brown under severe deficiency. The most striking symptom is the 'net veining' in the younger leaves where the leaf blade takes a striking yellow colour and the veins down to the finest branching stand out prominently dark green. At the advanced stage the bush is totally defoliated.

Calcium

Deficiency symptoms first appear in younger leaves and apices of plants. They do not grow to normal size, distort and crack, turn yellow and curl backwards. The maintenance leaves are affected slowly but defoliation of leaves and simultaneous death of bush may occur eventually.

Zinc

Leaves of zinc deficient plants turn yellow to dull olive green in colour showing the characteristics chlorosis and twist to a sickle shape and may develop wavy edge. Stunted growth, bunchy appearence of shoots, reduction in the distance of the internodes and leaf size are also sometimes noticed.

Copper

Copper of deficient plants have slightly darker leaves and very difficult to judge the difference in the field. Copper deficient leaves increases fermentation time than usual time in the factory and leaves become dark brown rather than bright orange.

Boron

Tip burning and dieback from the top are characteristics symptoms. Leaves may become narrow, twisted, leathery, crinckled and dark green in colour. There is cork formation on the petiole and on veins at the ventral surface of the leaves.

Manganese

The symptom is common in young tea. Chlorotic and necrotic spots in interveinal regions of leaves have been seen in the deficient plants. Sometimes leaves develop a mottling of red-brown spots and number of green veins reduced.

Iron

Deficiency results in interveinal chlorosis in younger leaves and whose veins remain green. When the deficiency is severe, scorching of the leaf tips and margins may occur.

Molybdenum

A bright green or pale orange colour of the leaves due to interveinal mottling, cupping of old leaves and necrosis at leaf tips and margins and drying out of leaves showing thin papery effect are the symptoms of deficiency.

Manuring

Manures are used in organic farming systems to feed the soil and the organisms living within it. These have no need of mineral fertilisers, but for organic matter which they can turn gradually into plant available nutrients. Satisfying the edaphone (the sum of all soil organisms) thereby also provides the optimum requirements for a sufficient and continuous supply of nutrients to the crops. This is achieved through the maximum production (depending on site conditions) of biomass per unit area.

The nutrients held in organic combinations become available to the tea plants slowly over many years as a result of natural control over nitrification effected by the heterotrophic organisms through nutrient immobilization, which is very beneficial to a perennial crop like tea.

Livestock and animal husbandry

Livestock play a major role in organic agriculture as the intermediary between the utilization of crop residues or fodder produced at the estate and returnof nutrients as manure. The number of livestock kept on a particular estate is therefore dependent upon the amount of fodder available, or the size of crop area used to grow fodder on the site itself. Planting fodder crops (especially legumes) is of particular help in improving the fertility of the soil. Hedges can be useful not only as windbreaks and as protection against erosion but they can also act as constant source of forage for cattle.

Industrialized large-scale livestock farming (many animals - not enough land) are not permitted in organic farming systems, because an optimal synergistic effect between the plants, soil and animals cannot be achieved with them. Strategies should be adapted according to the prevailing site and the way that animals are kept. For example, it is important to take care that over-grazing or damage through bites do not occur, as these could cause problems with erosion. If animals are to be introduced into a functioning farm unit, then care must be taken not only to choose the correct race of animal suited to local conditions, but also to make sure that the species itself can be adapted to the existing eco-system without causing any cakes, damage (www.naturland.de)

Cattle-rearing have no direct natural competition for food because they eat plants that are not consumed by humans; they produce protein in form of milk and meat, and also leather and wool; they can be used for transport purposes; they produce dung. Livestock keeping at farms is an age-old practice in many regions worldwide. Dairying in particular has helped a number of small and marginal planters to improve their income. Pigs can be easily adapted to a small farm structure, as they will feed on crop residues and organic waste products. Nevertheless, stalls must be provided with plenty of space for them to roam outside, to protect the environment - especially delicate eco-systems - against the animal's natural tendency to root around. Storage and application of cattle dung seldom attract proper attention of the farmers resulting in 40-60 percent losses in nutrients, especially nitrogen. Leaching of NO_3-N polluting the ground and surface water resources is usually observed from cattle dung pits. Organic farmers and farming methods take adequate care in minimizing these losses through adoption of technologies on composting, vermicompoisting etc. This not only improves the nutrient availability from organic resources but also prevent potential hazard of ground water pollution.

Animal husbandry in organic farming rejects in principle the concept of maximising a short-term performance of the animals. Instead, an attempt is made to achieve an optimum life performance of the animal. It is also important only to keep animals suited to the region that will be able to cope with the local climatic conditions, and are capable of withstanding the potential presence of diseases. Medication used in a preventive manner is forbidden in organic farming (with the exception of vaccinations).

The livestock farming significantly influence the production efficiency and has a great influence in the amount of compost prepared. However, in the mid country wet zone of Sri Lankan planters have experienced several difficulties in rearing livestock; the majority of them have given up the practice of rearing cows, which flourished one and a half decades ago. Livestock farming is practiced only by 32 percent of the sample farms of Sri Lanka. Farmers are unable to procure farmyard manure or compost from outside sources due to the uncertainty of contamination with inorganic matter, which affects their organic certification and involvement of high cost.

Composting

Composting is the microbiological conversion of biodegradable organic wastes to stable humus by indigenous flora, including bacteria, fungi and actinomycetes which are widely distributed in nature. In composting, however, such selective factors as carbon/nitrogen (C/N) ratio, moisture content, aeration, pH and temperature determine the prevalence and succession of microbial populations. Development of several compost production technologies like vermicomposting, microbe mediated, phospho composting, etc. improves the quality of compost through enrichment with nutrient bearing minerals and other additives.

In any biological system, healthy growth will always be able to withstand pest and diseases to an extent that it can put up its vigour. This holds good to the plant also. Well decomposed compost may provide several micronutrients compared to NPK fertilizer alone thereby providing the plant the required nutrients which would help to withstand pest and disease attack.

Tea estates which are not capable of providing sufficient amount of compost from organic materials produced on site are permitted to purchase certain organic materials from outside after obtaining approval by the certification body. These include; neem press castor cake, bone meal, coconut press cakes and cattle dung from extensive animal husbandry.

Box 7.1

Objectives of Composting

- Suppression of unpleasant odours
- Improvement of hygienic conditions
- Reduction of the germination capacity of weeds
- Maintenance and improvement of manurial value
- Increase of the biological activity of soils
- Positive influence of plant quality
- Minimum loss of nutrients during application
- Minimum additional investment expenditures
- Acceptable working conditions
- Minimum external energy requirements with regard to processing and use

Source: Ref. 7.

Selecting composting materials

The basic requirement for every type of composting method is the correct choice and preparation of the used compost materials. Factors that are important include the ratio between carbon and nitrogen (C:N ratio), the humidity and aeration. Decomposition will only take place to produce nutrient-rich compost when all of these factors are correct. The ratio between carbon and nitrogen in the materials chosen should ideally be 25-30:1, this may be a little higher at warm, humid sites. Composting will generally not take place or is extremely slow in an environment in which the C to N ratio is greater than 80. When wide C/N ratio materials such as cereal straw/plant residues and leaves are to be composted, mixing the mass with farmyard manure / slurry or nitrogen rich plant materials such as legume straw or green manure has been proved highly beneficial.

The carbon phosphorus ratio (C/P) should be 100:1 for proper microbial growth and digestion [8]. Solid wastes usually contain a sufficient amount of phosphorus for microbial growth.

Table 7.5. A selection of C: N ratios in materials usually used for composting.

Material	C : N- ratio
Saw dust	Up to 400
Maize stalks	50 - 150
Straw	50
Green material from legumes	20 - 30
Dung including bedding	20 - 25
Straw from legumes	15
Farmyard Manure	15
Top soil	10 - 12

Source: http://www.naturland.de

Rock phosphate (phosphorous compounds which have not been broken up, and are thus not readily available to plants) can be used by mixing it directly into the compost. During the composting process, and especially in an acidic medium, the rock phosphate is turned in part into forms that the plants can access. Furthermore, mycorrhiza fungi can also dissolve phosphates in the compost. The following points should be kept in mind before organic material is used from sources other than the farm itself:

- The origin of all organic material must be ascertained. Manure is only acceptable if it originates from organic or at least extensive animal husbandry and/or organic material if it is not polluted with pesticide residues or other contaminating substances.
- The importation (bought-in) of organic material into the farm system is limited, with the main proportion being produced on site.
- Every instance of organic material purchase both type and amount must be approved before hand by the certification body.

Selection of a compost site

Compost heap in principle must be protected against heavy sunlight (drying out) and too heavy rainfall (leaching). A suitable site would therefore be in the shade, under trees (Fig. 7.1). Further, method of transporting the organic material to the heap, and the distribution of the finished compost within the plantation should be kept in mind at the time of selecting the location of the compost site.

Construction of a compost heap

Maximum size of the heap should not be more than 1.75 m high and 2.4 m wide at the base. There is some flexibility in these guide measurements and the correct proportions will comewith experience, because they produce a roof-type of structure that helps to drain off any rainwater and thus prevent water-logging – especially in wet climates. In extremely dry regions, it can make sense to dig compost pits. However, the overall height of the heap will decrease with the decomposing process. There are several factors which are needed to be taken care for building a heap. These are:

- The organic material needs to be mixed by chopping it up (into small pieces, yet not too small) or layered alternately.
- Plant remains that are very woody need to be chopped to aid decomposition.
- The foundation needs to be well drained. If the ground tends towards water-logging, then the first layer should consist of loose material (e.g. branches).
- The compost heap should be protected from drying out and/or leaching by covering it with natural materials (e.g. straw or banana leaves) or with perforated foil (the perforations allow gases to escape and prevent internal sweating).

Fig. 7.1. Compost heaps in Oothu tea estate, South India

Regulating the rate of decomposition

Compost worms and other micro-organisms are necessary for the efficient decomposition of the compost material. The process can be accelerated by adding bits of old compost heaps. Special compost preparations can also be added which will support the process. The time required for the material to decompose depends upon the temperature, composition, carbon - nitrogen ratio, pH, moisture content, aeration etc.

Temperature

When organic wastes undergo decomposition, the insulating effect of the materials leads to conservation of heat and a marked rise in temperature takes place. Under normal composting condition with optimum moisture content and aeration, the entire composting process could be divided into four stages; mesophilic, thermophilic, cooling down and maturing. Temperature monitoring in a composting mass may indicate the amount of biochemical activity taking place. A drop in temperature could mean that the material needs to be aerated or moistened or that decomposition is in a late statge.

Disease carriers and weed seeds are killed off by the high temperatures inside the heap. The insides can attain a temperature of up to 80°C, which can then lead to a loss of N, and is therefore not desirable. The temperature should therefore not exceed 60°C. The inner temperature can quite easily be established; yet so-called compost thermometer is also sometimes used. If the temperature exceeds 60°C then removal of the covering, watering the heap are important.

pH

This is a measure of the acidity or alkalinity of the mixture. Decomposition will occur most readily in a neutral medium because most microorganisms grow best under neutral conditions. Under aerobic conditions a material that is initially neutral will experience a decrease in pH during the start of composting. This will normally be followed by an increase in pH which will result in a final state that is slightly alkaline. Most materials decomposing aerobically will stay within a pH range that is conducive for microbial growth thus eliminating any need for pH control [9].

Moisture

To achieve the optimal decomposition rate, the water content of the organic material should be between 50 and 60 percent by weight. If the moisture content is in excess of 60 percent, the compost becomes compact thereby reducing the amount of air present leading to anaerobic conditions. Moisture content at less than 45 percent is inadequate to satisfy the moisture needs of microorganisms. Low moisture content also lowers the temperature of entire compost mass. However, the required moisture has been found to range between

40 and 100 percent depending on the nature of organic material to be composted [10].

Aeration

To facilitate proper aerobic decomposition, an adequate supply of gaseous oxygen is necessary. Conversely, if little oxygen is available, anaerobic microorganisms will predominate and anaerobic decomposition will occur. Anaerobic decomposition is slower than aerobic decomposition, results in only a partial breakdown of organic matter, and often produces a disagreeable odour. Thus, aeration is important in composting techniques. Anaerobic conditions will only prevail if no oxygen is made available either through a lack of ventilation, compaction or if the compost heap were too wet that it would be said to be "waterlogged".

The factors like aeration and moisture can be counteracted manually by turning the heap, uncovering it or watering. A compost heap should be turned over around 3 times during its 3-6 month decomposition cycle. A basic rule states that the longer the decomposition takes, the more nutrients are also lost. Losses are caused by the gassing out of nitrogen or carbon compounds or by leaching of nutrient-rich draining water (e.g. potassium) (http://www.naturland.de).

Box 7.2

Standards for Organic Manure

Fertilizer Advisory Development and Information Network for Asia and Pacific (FADINAP), Bangkok have formulated the following standards for organic manure.

- pH (1 percent) : 6.5 - 7.5 more or less neutral
- Organic matter should not be less than 20.0 per cent
- Carbon : Nitrogen ratio (C : N ratio) -10:1 to 15:1
- Moisture content should not exceed 25.0 per cent
- Pathogens and heavy metals originate mainly from sewage sludge and urban garbage: absent

The above specifications are also applicable suitable for tea plantations.

In addition, electrical conductivity (EC) is also considered to judge their suitability for use in tea fields.

- EC (1 percent) up to 0.50 dSm^{-1} recommended for young tea.
- EC (1 percent) up to 1.5 dSm^{-1} recommended for mature tea only

Uses for compost

Finished compost has a crumbly structure. It smells pleasantly of earth, and contains only traces of rotting material. Large residual pieces can eventually be sieved out. Compost should be applied just before the main plucking times on the site.

Inoculation of bulky organic manures with culture of *Azotobacter* (10 kg ha^{-1}) and phosphate solubilising fungi (*Aspergillus awamori*) before application into the field is practiced in some tea estates. *Pseudomonas striata*, *P. Rathonis*, *Bacillus polymixa*, *Penicillium digitatum* and *Schwanniomyces occidentalis* are also identified as efficient phosphate solubilizers and these carrier-based phosphate solubilising microorganisms (PSMs) are recommended for tea [11].

In tea field where rock phosphate is applied, phosphate solubilising bacteria can render the benefit of solubilising the phosphate from applied rock phosphate, whereas the effect of vesicular arbuscular mycorrhizae (VAM) is more pronounced in the absorption and translocation of such solubilised phosphate into the plant system. An association of VAM with tea roots has been established and a higher influx of phosphate has been reported in tea plants in the nursery when their roots are inoculated with VAM [12].

Ramarethinam and coworker [11] described the method application of VAM and other related microbes to tea soils. Organic manure or compost could be used as the carrier material prior to incorporating VAM and related microbes. 5000 kg of organic manure / compost may be spread on a flat floor to a thickness of 6". Then sprinkle with water and expose to direct sun after covering with thin transparent polythene sheet. Sand bunding of 2-3" thickness may be made four sides of the sheet to secure for water seepage. However, if enough space is not available then sterilisation of the organic matter could also be done in smaller lots. For 5 mt of sterilised compost, incorporate 6 kg of VAM along with 6 kg each of *Trichoderma viridae* (an antagonistic fungus), nitrogen fixers (especially *Azospirillum*), phosphate solubilising bacteria and entomopathogens like *Beauveria bassiana* and *Paecilomyces lilacinus*. Adequately moisten the mixture and then heap in shade for about 7-8 days. The growth of *T. viridae* will be faster than the entomopathogens. Nitrogen fixers and phosphorus solubilisers will proliferate and on the other hand VAM will not. Therefore, if needed, the heap can be kept moist even for 15-20 days. But the addition of excess water should be avoided. After the desired period of time, heap can be spread out and applied to the fields. 5 mt compost will suffice for 2 hectares of tea. Application could be done when soil is moist which will ensure that the microbes remain alive and adapt themselves to a new environment, the tea rhizosphere. Broadcasting of the inoculated material just near the root zone should be avoided. After application, forking the soil in the drip circle where the feeder roots normally occur is preferred.

Incidentally, most of these microbes are approved for organic cultivation. Further, the products which are IMO certified should only be considered to ensure the quality and also pin the responsibility for the quality on the manufacturing / distributing organization(s).

Use of bio-fertilizers increases efficiency of applied nutrients and they are hence desirable. Biofertilizers like nitrogen fixing *Azospirillum* and phosphate solubilizing *Pseudomonas* and *Bacillus* spp. are capable of providing nitrogen and phosphate to the tea plants. Biofertilizers fortified with suitable bio-inoculants can be used in organic tea fields.

- It can give 20-50 kg nitrogen and 30-50 kg phosphate in tea for instant uptake.
- It liberates growth-promoting hormones, vitamins and helps to improve soil fertility.
- It suppresses incidence of pathogens, secret antibiotic to control disease.
- It increases tea yield by fixing nitrogen and solubilising phosphate.
- It reduces depletion of soil nutrient and provides sustainability of farming system etc.
- It is pollution-free and improves physical properties and health of the soil for better quality flavour.

Acid tolerant bio-formulations can be used @ 25 kg/ha. The material may be mixed with soil at 1:5 ratios and applied by shallow placement method in root zone of the bushes. However, certain biofertilizers used in the field crops are not suitable for application in the acidic tea soils. Therefore, it is necessary to identify acid tolerant strains of these bacteria and fungi for enriching the tea soils.

Vermiculture

The improvement of soil through vermiculture is also becoming popular and is an important part of organic farming to hasten the decomposition of organic residues (may see www.vermijadoo.com for details). It is the technique by which organic wastes, manure, sewage sludge and household wastes can be processed by indigenous species of earthworms into vermicompost (excreta or castings of earthworms). The average nutrient content of vermicompost is N, 1.6 percent; P_2O_5, 2.2 percent; K_2O, 0.67 percent; Ca, 0.99 percent and Mg, 0.15 percent. Of all creatures found in a fertile organic field, compost heap or farm, the earthworm is perhaps the most desirable. These creatures are untiring builders of soil fertility and great contributors to increase soil fertility and health. They are said to be the intestines of the earth. Earthworms could potentially be used in phytoremediation and phytomining of metals from soils. It has recently been reported in China that the microbial populations, and mobility and bioavailability of heavy metals like Zn, Cu, Cr, Cd, Co, Ni, and Pb increased due to earthworm activity [13]. Dissolved organic carbon in soils is likely to be increased by earthworms through decomposition of organic matter into smaller water soluble compounds.

Preparation of vermicompost

Lampito mauritii, *Lumbricus rubellus* and/or *Perionyx excavator* worms can be used. All species of earthworm are not active in soil formation. *Eisenia foetida* for example cannot survive over a long period or thrive in field soils but normally found near manure dumps. Similarly, *Pheretima elongate*, *Lumbricus terrestris*, *Allolobophora caliginosa* do not tolerate culture conditions. A shallow wooden box of 45 cm x 65 cm x 20 cm with drainage holes may be used for rearing and multiplication of worms. This can be stored on shelves and tiers. The material for bedding is compounded for miscellaneous organic residues viz. rice husk, cereal straw, saw dust etc and is well moistened with water. The same is covered with wet bag, stored for about one month and is thoroughly mixed. When fermentation is over, the green manure such as water hyacinth (*Eichhornia* sp) or Guatemala grass (*Tripsacum laxum*) and poultry manure (without contaminating substances) may be mixed up. The material should be sufficiently loose when placed in the boxes

Fig. 7.2. Earthworms collected from vermicompost bin.

to facilitate worms to burrow and retain moisture. According to the nature of the material the proportion may vary but the protein content of the material should be about 15 percent, pH as neutral as possible and the boxes should be kept at temperature within 20 ^{0}C to 27 ^{0}C. 100 grams worm obtained from breeders stock are required for each 0.1 m^2 of surface area. The worms are regularly fed vegetable wastes, poultry manure and green manure @ 1 kg per day for every kg of worms [14].

Composting

The most convenient dimension of pit is 2 metre long, 1 m wide and 1 m deep. Pits are dug with sloping side. It can be made of brick or mortar. A layer of small pieces of bricks and sand should be used to form the base of the pit. It provides in draining the excess water if any in the pit. The pits should be then lined with old gunny bags to keep the worms from escaping into the surrounding soil. The pits can be filled with farmyard manure, green weeds, waste leaves, saw dust, tea leaves, and leafy biomass as available, and other organic residues. Cow dung and leafy biomass need to be used @ 60:40 or 50:50 ratios. A layer of 15 cm loamy soils should be sprinkled and kept moist for a week or so.

To harvest the worms from the boxes, two-third worms should be picked out from the box and placed on the top of the heap and should be kept sufficiently moist. The original box can be provided with fresh bedding material and the remaining worms will again multiply. The compost pits are to be left for a period of around two months and shaded. The pits must be kept constantly moist but never flooded. It has been observed that 10 kg of castings per kg of worms are generally produced.

Box 7.3

Investment and Return from Vermicomposting

Investment		
Construction of thatched/shaded-2 tanks of size 20ft x 4 ft x 2 ft	Rs.	7,500=00
Cost of worms	Rs.	2,000=00
Miscellaneous	Rs.	1,000=00
Transportation	Rs.	1,500=00
Total	Rs.	12,000=00
Return		
1st year 4 tons vermicompost	Rs.	8,000=00
2nd year onwards--5 tons vermicompost	Rs.	10,000=00

* Source: Ref. 15.

It has been reported that the duration of times during transition from conventional agriculture to organic agriculture could be curtailed to three months by harnessing and seeding the soil with vermicastings [16]. The role of earthworms in improving soil productivity to get more yields with lesser irrigation and lowering the cost of production is well established but has not pervaded the sections of the farming community including tea planters for efficient farm application.

Vermicompost is mixed in nursery soil while filling up of polythene sleeves for good rooting of the tea cuttings. This can be also mixed with the soil at the time of infilling and replanting for easy establishment of plants. The compost can be also used in mature tea fields for enrichment of poor soil organic matter and enhancement of the productivity of soil. For having proper amount of potash nutrient in the vermicompost some quantity of banana leaf and *Alocassia* spp (Mankachu in Bengali) leaf is required to be used as raw material. During replanting or new planting of tea, vermicasting produced by the burrowing earthworms may be used. Compost @ 5 tonnes ha^{-1} may be applied in presence of sufficient moisture. Vermiculture may not be successful in conventional tea fields for the following reasons [17].

- Chemical weed control is widely practiced.
- Soil reaction is extremely acidic to very strongly acidic.
- Soil organic matter in tea fields is generally low to medium.
- Soil remains too wet in rainy season.
- Soil temperature sometimes goes below 10 ^0C during winter months owing to use of various inorganic fertilizers for nutrition of tea.

Application dose for organic manure

An idea of the general application of NPK in major tea-growing areas of India can be had from Table 7.6.

Table 7.6. General recommendation of NPK in different tea-growing areas of India.

Darjeeling hills						
Yield range (kg) Made tea/ha annual cycle	**N**	**P_2O_5**	**K_2O Available K in the soil**			**Equivalent dose of organic of manure (kg)**
			60	**60-100 (ppm)**	**100**	
Up to 1000-2400	22.5-80	9-12	4.8-6	4.5-6	45.6	1125-4000
Up to 600-1000	15-22.5	6-9	3-4.5	3-4.5	3-4.5	750-1125
Up to 600	15	6	3	3	3	750

North East India.						
Yield range	**N**	**P_2O_5**	**K_2O kg ha^{-1} at available K level (ppm)**			**Equivalent organic manure mt ha^{-1}**
Processed tea (kg ha^{-1})	**(kg ha^{-1})**	**(kg ha^{-1})**	**<60**	**60 – 100**	**> 100**	
(Cycle average)						
Up to 1500	up to 90	20	up to 90	up to 70	up to 50	4.5
1500 – 2000	90 – 110	20 – 30	90 – 110	70 – 80	50 – 70	4.5 – 5.5
2000 – 2500	110 – 140	30 – 50	110 – 140	80 – 120	70 – 100	5.5 - 7
2500 – 3000	140 – 165	50	140 – 165	120 – 140	100 – 120	7 – 8.2

South India				
	Organic manure (kg/ha)			**Equivalent dose of organic manure (kg)**
Year	**N**	**P_2O_5**	**K_2O**	
I	49	9	4.5	1225
II	60	12	7.5	1500
III	75	15	7.5	1875

Fertilizer schedule g/plant (in two splits) for young tea				
Age	**N**	**P_2O_5**	**K_2O**	**Equivalent dose of organic manure (g)**
1st year	5	2	1	250
2nd year	7.5	3	1.5	375
3rd year	15	6	3	700
4th year	25	10	5	1250
5th year	50	20	10	2500

Manuring for organic tea in pruned year (for 2500 - 4000 kg made tea/ha)

Manure	Quantity	Nutrients (Kg/ha)		
		N	P	K
1.Compost*	10 t/ha	199	14	72
2.Neem cake ** (1 split)	2.5 t/ha	121	9	26
3.Castor cake *** (1 split)	2.5 t/ha	121	9	28
4.Rock phosphate (20% P_2O_5 (1 split)	400 kg/ha	-	35	-

Manuring for organic tea in other years (for 2500 - 4000 kg made tea/ha)

Manure	Quantity	Nutrients (Kg/ha)		
		N	P	K
1.Neem cake ** (2 splits)	5 t/ha	260	18	56
3.Castor cake *** (2 splits)	5 t/ha	280	18	56
3. Wood ash 5.5% K (1 split)	2.5 t/ha	-	-	137
4.Rock phosphate (20% P_2O_5 (1 split))	400 kg/ha	-	80	-

* Compost (1.99 % N, 0.14 % P, 0.72 % K, moisture less than 10.0 %); ** Neem cake (4.85 % N, 0.36 % P, 1.03 % K); ***Castor cake (4.82 % N, 0.35 % P, 1.11 % K)

Rock phosphate may be applied in the pruned and third years as placement based on soil test values. Soil available P should be maintained at and above 22 ppm.

Wood ash may be applied only in mature tea fields preferably in dry periods by broadcasting 15 cm away from the collar.

Source: Ref. 18, 19 and www.upasitearesearch.org

For high-yielding sections suchas clonal blocks, additionally two tons of compost and 2.5 tons of oil-free castor cake or neem cake may be applied. In case of dearth of compost, five tons of compost and two tons of oil-free neem cake or castor cake per hectare in two splits as one in pre-monsoon and one post-monsoon may be applied. Organic manure should preferably be applied in the staggered trenches. The dimensions of trench should be 2 x 0.3 x 0.45 m and 2 m between two trenches. Trenches may be taken across the slope ones in every two or three rows depending on the gradient [19].

Requirement of potassium is usually met by the addition of wood ash. But in high-yielding fields potash deficiency has been noticed and this could to be rectified by the application of a mined product like langbeinite or patent kali which contains potassium as well as magnesium. Requirement of phosphate can be met by the application of rock phosphate. Basic slag, natural rock phosphate, limestone, chalk, gypsum etc are used occasionally with the prior approval of the inspection body. Zinc is only the micro-nutrient required by tea in large quantity and its deficiency is very much noticed in almost all organic tea fields.

The general recommendation for NPK should be based on the intrinsic needs of the plants and their productive potentials under an organic regime. The recommendation may vary with locations and eco-specific conditions, and cannot be generalized. The major nutrients like nitrogen, potash and phosphate will have to be sourced to the natural products, in addition to those from compost, mulch, legumes and other permitted inputs.

Biodynamic agriculture

In the early twenties some group of farmers in Germany were deeply concerned with the decline in the soil fertility, loss in the food quality and the uneconomical animal husbandry. They approached Dr. Rudolf Steiner, philosopher to get rid of these problems. In 1924, Dr. Rudolf Steiner gave the solutions as 'biodynamic agriculture' in the form of eight lectures at Koberwitz. His knowledge and thinking was much influenced by India's Vedas. In this farming, the farm is considered as 'living organisms' in which one activity influences and affects the other [20]. Biodynamic agriculture can otherwise also be called as a 'Holistic Farming System'. It does not require sophisticated facilities and most of manures and biopesticides are prepared in the farm. Biodynamic preparations are components of biological agriculture, capable of affording long-term sustainability to agriculture and particularly to the ecosystem. Basic principles of biodynamic farming are to restore the soil and the organic matter in the form of humus, increasing microbial population, skilful application of the factors contributing to soil life and health, treating manure and composts in biodynamic way. This farming system can be practised in any organic farming system. The cow is considered as central animal in biodynamic agriculture and its horn also plays an important role.

Biodynamic agriculture has been making waves over the past decade in India, with the system slowly, but surely, winning converts as an ethical alternative to conventional farming methods. Interestingly, some progressive tea estates brought in internationally renowned organic and biodynamic consultants to advise them. While Tadeu Caldas of Ecotropic (www.ecotropic.com) assisted Ambootia and BBTC, Peter Proctor of Biodynamic Outreach, New Zealand is an inspiration at the training programmes of the Bio-Dynamic Association of India (BDAI), an organization which gains new converts through each course. Currently, several South Asian consultants are active in the field, including Natura Agrotechnologies (www.naturaagrotechnologies.com) that uses techniques that bridge biodynamic and Vedic agricultural processes. Recently, Swiss and German buyers have shown considerable interest in procuring biodynamic teas. In biodynamic farming Dr. Rudolf Steiner, philosopher, has given eight recipes as 'biodynamic preparations'. They are as follows:

Biodynamic preparations

Horn manure (BD 500)

Straw-free dung of a lactating cow was filled into uncovered cow horns and buried in the ground during autumn and winter months (Sept-Oct. in India) (Fig. 7.3). The cow horns can be procured from slaughter houses. The horns are identified by being solid at the top and have characteristic lactating rings at the base (Fig. 7.4). The dung inside the horn undergoes breakdown process due to various microbial activities during the autumn and winter months. The horns containing dung are dug out in the spring (Feb-March in India). The dung gets totally transformed into humus. Soon after removing the manure from the horn, it must be protected from drying out by storing in suitable glass or ceramic vessel or mud pot covering by wet peat or moss.

Box 7.4

Marketing and Distribution of Biodynamic Produce

Producers, processors and traders can be granted the right to market under the DEMETER and Biodyn trademarks by entering into annually renewable contracts with the Demeter Guild. Binding regulations for management and purchased inputs are issued by this organization. Usually two persons, the inspector or advisor and a local planter, are assigned to control adherence to the regulations. The Guild enjoys the confidence of the consumer circles.

As a result of the activities of the Demeter Guild around the world, there is a well-established market for biodynamic produce, especially processed products, babyfoods and teas, which allow producers to make a respectable living.

The details of the Demeter and Biodyn standards, is available from the Demeter Bund, e. V., Germany, Bio-Dynamic Agricultural Association in Britain and its related organizations around the world.

Fig. 7.3. Straw-free manure is filled into cow horns which are buried in the ground of Ambootia tea estate.

Fig. 7.4. Horns have characteristic lactating rings at the base.

Application rate and stirring: One set (35 gram) of preparation of BD 500 in 13.5 liters of water for 0.4 ha (one acre) of soil can be applied. This amount of preparation BD 500 in 13 liters of water for 0.4 ha is stirred in a bucket of 15 liters capacity. However, sufficiently large barrels have to be used for spray. For example to spray 10 ha, barrel of 400 liters capacity should be used. Stirring is done with the help of a wooden stick and is stirred in one direction until a funnel (vortex) is created in water reaching the bottom of the container. One such vortex is obtained, the direction is abruptly changed. Again a vortex is created and the direction of the vortex is changed again and so on. Likewise stirring is carried out for one hour.

The solution would be energized after one hour stirring and is sprayed on to the soil with a knapsack sprayer or by hand application using a brush or a broom during evening hours.

Horn silica (BD 501)

This silica spray is based on a very finely ground quartz, or a silicate. The quartz is powdered (Fig. 7.5) by crushing it with a pounding stone, a mortar and pestle and finally ground between glass plates. The powder is made into slurry by adding water and packed into the horns and buried during the spring and summer months (March-April). The horns are dug out of the ground in Sept- Oct and stored in transparent glass containers near the window sill.

Fig.7.5. Ground quartz is put into cows horn and buried in the soil.

Application rate and stirring: One set (one gram) of preparation of BD 501 in 13.5 liters of water for 0.4 ha of soil can be applied. This amount of preparation BD 501 in 13 liters of water for 0.4 ha is stirred in a bucket of 15 liters capacity. Stirring should be done as followed in BD 500. Spray as fine mist over the canopy of tea bushes early in the morning during the ascending period of the month. For this purpose the sprayers should be used by fitting it with NMD or suitable nozzle, in high pressure. The plant experiences shocks and is detrimental to crop growth if spray is done during hot sunshine hours.

Yarrow preparation (BD 502)

Sheath: Bladder of a male deer

Contents: Fresh or dried yarrow (*Achillea millefolium*) flowers (Fig. 7.6), in full bloom, collected on a sunny day.

Procedure: Fill the dried bladder with flowers and bind closed. Hang the bladder in the sun before mid-summer (Fig. 7.7), if necessary protected from the birds with wire netting. Bury it in the soil in the

Fig. 7.6. Yarrow plant with flowers.

Fig. 7.7. Yarrow flowers in the bladder of a deer in the sun.

early autumn, and recover it between end of March and April at the latest.

Camomile preparation (BD 503)

Sheath: Intestine of a cow.

Contents: Fresh or dried camomile (*Malricarja chamomilla*) flowers (Fig. 7.8) collected in the early morning.

Fig. 7.8. Camomile flowers.

Procedure: The intestine is filled with the camomile flowers, the intestine allowed to dry a little, and in early autumn buried in the soil. It is recovered before the end of April.

Nettle preparation (BD 504)

Sheath: Sack or wooden box.

Contents: Stinging nettle plants (*Urtica dioica*) cut in the early morning at the start of flowering.

Procedure: The wilted nettle is bundled together, and packed into the sack or box. It is buried in the soil at mid-summer, surrounded by a layer of peat five centimetres thick, and recovered one year later.

Oak bark preparation (BD 505)

Sheath: Skull of an animal.

Contents: Finely ground bark from oak (*Quercus robur*), taken from the trunk with a plane in the early autumn.

Procedure: The oak bark powder is filled into the skull through the foramen magnum (the hole at the base of the skull through which spinal cord entered) packed firmly, and the hole closed with a piece of bone held in place with clay. It is placed into a drum of rainwater, or at the edge of a pond, in the presence of decaying plant material and in a flow of water, in the early autumn. In the spring it is taken out, the contents removed and dried.

Dandelion preparation (BD 506)

Sheath: Mesentery of a cow.

Contents: Dandelion (*Tarxacum officinale*) flowers (Fig. 7.9) in the early stage of blooming collected on a sunny morning in spring.

Procedure: Moisten the dried flowers in autumn with dandelion tea, wrap up in the mesentery, and tie with a string (Fig. 7.10). Bury in the autumn and recover in the spring. It separates the metabolic from the respiratory system of the heap.

Fig. 7.9. Dandelion plant with flower.

Fig. 7.10. The dried dandelion flowers are wrapped in the mesentery which is tied together with a string.

Valerian preparation (BD 507)

The fresh valerian (*Valeriana officinalis*) flowers (Fig. 7.11) are pressed, and the juice put into bottles. Leave them uncorked for six weeks to allow fermentation to finish, then cork and store in a dark cellar. It acts as the thermostat of the heap, regulating its temperature.

Equisetum preparation (BD 508)

Dried shoots *of Equisetum* - a tea is made in a vat, covered and filled with rain water. The tea is sprinkled, which acts as the immune system of the compost heap.

Fig. 7.11. Valerian plant.

After recovery the preparations 502 and 506 are removed from their sheaths, and stored separately in glass, ceramic or stoneware containers, which are placed in a box and each container is surrounded on all sides with peat (Fig. 7.12). This preparation box should be kept in a cool, dark, frost-free place.

Fig. 7.12. Ceramic containers is surrounded on all sides with peat in Makaibari tea estate.

The biodynamic preparations are used for enhancing maintaining soil fertility and increased photosynthetic activity leading to a superior crop. It also increases resistance to plant diseases [21]. Banerjee and coworker [18] listed out the benefits of spraying BD 500 preparation which are as follows.

- Increase earthwork and microbial activity.
- Increases availability of insoluble minerals.
- Improves soil structure and moisture-holding capacity.
- Increase in plant quality.

The yarrow flowers (BD 502) bear a relationship to sulphur and potassium. This preparation permits plants to attract trace elements in extremely dilute quantities.Camomile (BD 503) assimilates calcium and potassium. The stinging nettle (BD 504) bears a relationship to sulphur, potassium, calcium and iron. It stimulates soil health and is said to enliven the soil. The ash of the oak bark contains up to 76 percent calcium. The moon forces would be said to become active in the soil in a healthy normal manner through prep 505, the oak bark. It provides healing forces (qualities) and is an aid in combating plant diseases [22].

Biodynamic compost preparations (BD 502-507) are called BD set. BD sets are used in the Cow Pat Pit (CPP), BD compost, BD liquid manure and biodynamic liquid pesticides.

Cow Pat Pit

Cow Pat Pit or CPP refers to manure which is made out of cow dung, egg shells and basalt and is used also in small quantity. CPP manure enhances rooting of plant cuttings. Besides rectifying micro-nutrient deficiency in plants and soil, CPP manure can also be applied for various other farm requirements.

Materials required for CPP preparation:

- Fresh cow dung (60 kg) is collected from a lactatine cow
- Crushed egg shells (as a source of calcium)
- Borewell soil/basal rock/blue granite dust 100 g (as a source of silicon dioxide and minerals)
- Two sets (2 g each) of BD 502-506 and 20 ml of BD 507 solution (4 percent)

Procedure of preparation:

- Dig a pit and line the four sides of the pit with baked bricks. The size of the pits should be 3 x 2 x 1 feet dimension. The bottom of the pit should be lined.
- A site with good drained not prune to dry out shoud be selected.
- Sprinkle water to turn the cow dung into dough like consistency.
- Crushed egg shells and minerals mentioined above are sprinkled over the cow dung.
- The bricks lining is moistened with water before placing the dung mixture.
- Mix the dung for 15 minutes and place it in the pit. Dung should not be tightly packed.
- Insert two sets of 502-506 (2 g each) preparations by making holes over the dung layer. Two sets of 507 preparations (20 ml of 4 percent) is stirred for 15 minutes in 2 litres of water and sprinkled over the cow dung.
- To maintain moisture, a wet jute sack over the CPP may be placed. The pit should be in a well-shaded and aerated area.

- After one month the manure is gently aerated with a fork for uniform mixing.
- CPP manure matures in 2-3 months and it can be stored in an earthen pot placed in a dark room, avoid the CPP from drying.

Application rate and stirring : CPP can be applied along with BD 500 preparation. After 45 minutes of stirring add 1 kg of CPP to the solution and stir both BD 500 and CPP for another 15 minutes. The spray solution is then sprayed over the soil of one hectare area.

Liquid Manure

A large number of liquid manure is made out of locally available plant materials which are diluted and used regularly as a tonic for the crops. Animal base liquid manure comprises cattle dung, sheep and goat droppings, chicken manure, fish manure etc, which have not been treated by allopathic medicines or antibiotics can also be used. The liquid manure has to be diluted with water prior to application (1:10).

In general, it is observed that organic tea planters lack integrated management of all fertilizing possibilities available to them, such as the establishment of contour hedgerows, efficient shade plant management, optimum use of crop residues, mulching, use of legumes as cover crops, manual weeding and integrated farming, which help to develop fertilizing self sufficiency through maximum use of plant and animal resources. Therefore, in principle, necessary knowledge and training on sound organic farming practices that improve soil fertility for economic advantages should be provided to planters to encourage higher efficiency levels.

References

1. Rai, M. 2005. Organic farming: potential and strategies. *14th Millennium Guest Lecture at S. V. Agricultural College*, 3rd June, 2005, Tirupati, Tamil Nadu, India, 1-13.
2. Gaur, A. C. 1994. Bulky organic manures and crop residues. In: *Fertilizers, Organic Manures Recyclable Waste and Biofertilisers* (ed. H. L. S. Tandon). Fertiliser Development and Consultation Organisation, New Delhi, 36-51.
3. Chakravartee, J. and Sinha, M. P. 1994. Soil management in tea for north east India. In: *Field Management in Tea* (ed. J. Chakravartee), Tea Research Association, Tocklai Experimental Station, Jorhat, India, 34-40.

4. Banerjee, B. 1993. *Tea Production and Processing*. Oxford & IBH Publishing Co., Pvt. Ltd., New Delhi, p.336.
5. Konishi, S., Miyamoto, S. and Tazi, T. 1985. Stimulatory effects of aluminium on tea plants grown under low and high phosphorus supply. *Soil Sci. Plant Nutr.* **31**(3): 361-368.
6. Tandon, H. L. S. and Ranganathan, V. 1993. Fertilizers and their management in plantation crops. In: *Fertilizer Management in Plantation Crops: A Guide Book.* (ed. H. L. S. Tandon), Fertilizer Development and Consultation Organisation, New Delhi, 26-60.
7. Vogtmann, H. and Besson, J. M. 1978. European composting methods: treatment and use of farmyard manure and slurry. Compost Science/ Land Utilisation. Reprinted In: *EFRC (1980). The Research Needs of Biological Agriculture in Great Britain.* Elm Farm Research Centre Report No. 1.
8. Hagerty, D. J., Pavoni, J. L. and Heer, J. E. (Jr). 1973. Solid waste management, Von Nostrand Reinhold, New York.
9. Eberhardt, D. L. and Pipes, W. O. 1974. Illinois composting application. In: *Large scale composting*. (ed. M. J. Satriana). Noyes Data Corporation, Park Ridge, New Jersey.
10. Gaur, A. C. and Sadasivam, K. V. 1993. Theory and practical considerations of composting organic wastes. In: *Organics in Soil health and Crop production* (ed. P. K. Thampan). Peekay Tree Crop Development Foundation, Cochin, Kerala, India, 1-22.
11. Ramarethinam, S. and Murugesan, N. V. 2005. Some aspects of arbuscular mycorrhizal biofertilizer and phosphorus solubilising microorganism. *Planters' Chronicle* **101**(4): 17-20.
12. Rajagopal, B. and Ramarethinam, S. 1997. Interaction between vasicular arbuscular micorrhizae and tea root. *Planters' Chronicle* **92**: 263-268.
13. Wen, B., Hu, X-Y., Liu, Y., Wang, W-S., Feng, M-H. and Shan, X-Q. 2004. The role of earthworms (*Eisenia fetida*) in influencing bioavailability of heavy metals in soils. *Bio. Ferl. Soils* **40**: 181-187.
14. Ghosh Hajra, N. 1994. Vermiculture and vermicasting in tea cultivation. *Assam Rev. Tea News* **83**(5): 9-12.
15. Anonymous, 2005. Ecological and Economic Sustainability of Tea Gardens in Darjeeling District of West Bengal – A Feasibility Report (Draft Revised Final Report) – by CCBA under West Bengal Forest Development Corporation Ltd., India, p.131.
16. Bhawalkar, U. S. and Bhawalkar, V. U. 1993. Vermiculture biotechnology. In *Organics in Soil Health and Crop Production* (ed. P. K. Thampan). 69-95.
17. Verma, D. P. 1993. *UPASI TRI Tea Newslet.* **2**(2): 6.
18. Banerjee, B. and Chaudhuri, T. C. 2001. Package and practices for production of organic tea in India. Tea Board, *NTRF*, p. 36.

19. Radhakrishnan, B. 2004. Package of practices for organic tea. *Planters' Chronicle* **100**(1): 19-31.
20. Jerome, J. J. 2003. Biodynamic farming system. *Assam Rev. Tea News* **91**(12): 24-29.
21. Caldas, T. 1997. Soil fertility improvement and plant nutrition strategies in organic and biodynamic tea production: A basis for sustained yields, cup quality, pest resistance and sustainability. *Proc. 2nd Inl. IFOAM Conf.* April 28 -29, 1997, Calcutta, India, pp. 49-51.
22. Rosenberg, A. and Linders, T. 2004. *Organic Agriculture: A Handbook.* Lindros Whole Earth Consultants, South Africa, p.221.

Chapter 8

Pest Management

Tea plant is an evergreen woody perennial provides suitable condition for the pests to feed on it and breed. The most common pests of tea are mites, thrips, tea mosquito bug, leaf-eating caterpillars, mirids, scolytid beetles and termites. It was estimated that more than 1000 species of arthropods and 82 species of nematodes are known to attack tea all over the world [1], though only about 300 species of insects and mites are recorded from India. Nearly 400 pathogens are associated with tea globally but only 58 pathogenic fungi are known to cause diseases in tea in India. Further, more than 130 species of weeds cause problems in young tea clearings, pruned fields and vacant patches of tea gardens in India.

Since the magnitude of pest infestation varies seasonally depending on altitude, climate and cultural practices, an exact assessment of crop loss is very difficult. However, it has been estimated that the crop loss is accounted for anywhere between 6 and 14 percent due to the ravages of insect pests [2], and 10 and 15 percent due to diseases [3].

International regulations stipulate that pests and diseases should be primarily tackled by the planting of tolerant or resistant cultivars, manipulation of microclimate (such as through shade regulation) and pruning etc., and use of eco-friendly approaches like biological control. However, certain products are allowed for use in organic agriculture for the control of pests and diseases in plant production. Such products shall only be used when absolutely necessary and should be chosen taking the environment into consideration. Many of these products are restricted for use in organic production (Table 8.1).

Field scouting by the supervisors/managers as and when necessary should be done to monitor the initial development of pests and diseases in each section of the tea estate. Based on the findings of field monitoring and if pest infestation attain the economic threshold level (see Box 8.1) then appropriate plant protection measures should immediately be taken.

Box 8.1

Economic Threshold Levels (ETL) of Pests

Based on field studies on crop loss due to pests, the ETLs for important tea pests and diseases have been worked out as follows:

- Tea mosquito bug (5 percent infestation);
- Thrips (three thrips per shoot);
- Shot hole borer (15 percent infestation);
- Flushworms, tortrix (5 infested rolls and tea tortrix per bush);
- Red spider mites (4 mites per leaf);
- Pink and purple mites (5 mites per leaf);
- Blister blight (35 percent infection);
- Grey blight (18 percent infection);

It should be noted that ETL depends not only on the quantum of crop loss, but also on the market price of tea.

Source: Ref. 4.

Natural enemies

Pest and disease control in organic tea cultivation is primarily preventive rather than curative. The basis of insect, disease and nematode management in organic farming systems is the reliance on the inherent equilibrium in nature. The ecosystem in tea fields is complex and stable, and the effects of natural enemies on the population of insect pests are substantial even under insecticidal control. Since excessive insecticide application reduces the effects of natural enemies and causes the resurgence of insect pests, conservation and utilization of indigenous natural enemies to the highest extent is essential in the control system of tea pests. Natural enemies of insect pests can be divided into a few large groups: predators, parasitoids, pathogens, and nematode. Some of their main characteristics are listed in Table 8.2.

A considerable number of natural enemies of pests of tea have been identified (Table 8.3) and some of them have been proved capable under ideal conditions of keeping the pest population below economic threshold levels. A single ladybird beetle can devour an entire colony of thrips or mites in a day. Release of lady birds in badly infected areas has given excellent results and the infested areas have shown almost complete recovery within 45 days followed by prolonged mite-free period [5]. A coccinellid predator, *Stethorus gilvifrons* was found to predate on red spider mite consuming 10-12 mites per day [6]. Phytophagous mites infesting tea are mainly preyed upon by several predatory mites. The pink mite (*Amblyseius herbicolus)* and *Euseius ovalis* are the two common phytoseiids feeding on the eriophyids, the pink mite (*Acaphylla theae)* and the scarlet mite (*Calacarus carinatus*). The stigmaeid, *Agistemus flescneri* is an important predator of eggs and nymphs of *Oligonychus coffeae* in the north east India [7]. Mites belonging to *Pronematus, Parapronematus* and *Tydeius* are also active in tea fields. *A. herbicolus* is the most common predator of the mite pests. Population of this predator reaches the peak during December and February in South India and a single predatory mite could consume more than 400 pink mites during its life span [8].

Table 8.1. Products for use in pest management.

Substances description, compositional requirements	Conditions for use
Plant and animal origin	
* Algal preparation	
* Animal preparations and oils	Restricted
* Beeswax	Restricted
* Biodynamic preparations	
* Chitin nematicides (natural origin)	
* Coffee grounds	
* Corn gluten meal (weed control)	
* Dairy products (e.g. milk, casein)	
* Ethyl alcohol	Restrcted
* Fungal preparations	
* Gelatine	Restricted
* Lacithin	
* Light mineral oils (e.g. paraffin)	
* Lime sulphur (calcium polysulfide)	Restricted
* Natural acid (e.g. vinegar)	
* Neem (*Azadirachta indica*)	Restricted
* Pheromone traps and dispensers only	
* Physical methods (e.g. chromatic traps, mechanical traps)	
* Plastic mulches	
* Plant oils	
* Plant preparations	Restricted

Substances description, compositional requirements	Conditions for use
* Plant based repellents	Restricted
* Potassium bicarbonate	Restricted
* Potassium permanganate	
* Propolis	
* Pyrethrum (*Chrysanthemum cinerariaefolium*)	Restricted
* Quassia (*Quassia amara*)	Restricted
* Quicklime	Restricted
* Release of parasites, predators and sterilized insects	Restricted
* Ryania (*Ryania speciosa*)	Restricted
* Seasalt and salty water	Restricted
* Tobacco tea (pure nicotine is forbidden)	Restricted
* Sabadilla	
* Rotenone (*Derris elliptica, Lonchocarpus* spp., *Thephrosia* spp.)	
Mineral origin	
* Chloride lime	
* Clay (e.g. bentonite, perlite, vermiculite, zeolite etc)	
* Copper salt (e.g. sulphate, hydroxide, oxychloride, extenuate)	
* Diatomaceous earth, light mineral oils (paraffin)	
* Lime sulphur (calcium polysulphide), soft soap	
* Silicates (e.g. sodium silicates), quartzs, sodium bicarbonate	
* Sulphur dioxide	Restricted
* Sulphur	Restricted
Microorganisms	
* Fungal preparations	
* Bacterial preparations (e.g. *Bacillus thuringiensis*)	
* Viral preparations	
Others	
* Calcium hydroxide, calcium dioxide,	
* Homeopathic and ayurvedic preparations	
* Soda	

'Restricted' means that the conditions and the procedure for use shall be sent by the accredited certification programme.

Table 8.2. Characteristics of natural enemies of insect pests.

Predators	• Predators usually hunt or set traps to catch a prey to feed on. • Predators can feed on many different species of insects. • Both adults and young insects (larvae or nymphs) can be predators. • Predators follow the insect population by laying more eggs when there is more prey available. • Common predators are spiders, lady beetles, ground beetles, syrphid flies, and large stinging wasps.
Parasitoids	• Attack only one insect species or a few closely related species. • Only the larvae are parasitic. One or more parasitoid larvae develop on or inside a single insect host. • Parasitoids are often smaller than their host. • Common parasitoids are tiny wasps or flies.
Pathogens	• Insect pathogens are fungi, bacteria or viruses that can infect and kill insects. • Pathogens require specific conditions (e.g. high humidity, low sunlight) to infect insects and to multiply. • Most insect-pathogens are specific to certain insects groups, or even restricted to certain life stages of an insect. • Commonly used insect pathogens are *Bacillus thuringiensis* (Bt), and NP virus.
Nematodes	• Nematodes are tiny worms (about 1 mm long) that naturally live in soil. • Entomopathogenic nematodes attack and kill insects, and do not harm plants. • Entomopathogenic nematodes are usually only effective against pests in the soil, or in humid conditions

Source: Ref. 9.

Spider is another group of natural enemies prey on moths, leafhoppers, aphids, bugs etc with strong predacious ability [10].

The green lacewing *Chrysopa madestes* and the staphylinid *Oligota pygmaea* are also the major predator. In South India, predatory green lacewing *Mallada boninensis* (Okamoto) feeding on eggs, nymphs and larvae of red spider mite was reported recently [11]. Good control (80-90 percent) of red spider mite and kanazawai spider mite, *Tetranychus kanazawai* was also achieved when larvae of green lacewnig *Mallada basalis* (Walker) is released to tea plantation in Taiwan [12].

The predatory thrips *Aeolothrips intermedius* and *Mymarothrips garuda* are important natural enemies of thrips [7]. Recently in North East India a neuropteran pedator, *Chrysoperla carnea* was found effective against thrips and *Helopeltis* @ 500-100 adults/ha (www.bioagrosolutions.com).

Table 8.3. Important natural enemies of some tea pests.

Pest	Natural enemy
Red spider mite *Oligonychus coffeae* (Nietner)	*Verania vincta* Garh., *Stethorus qilvifrons* Mulsant, *Scymnus* sp., *Jaurevia quadrinotata* Kapur, *J. soror, Verticillium lecanii* (fungus), *Agistemus* sp., *Chrysopa* sp., *Phytoseilus persimilis* (Exotic sp.), *Chrysoperia carnea* Stephens, *Exothoris caudata, Pronematus* sp., *Cunaxa* sp., *Amblysieus herbicolus* (Chant), *A. ovalis, A. mcmurtry*, Staphilinid beetle.
Pink and purple mites, *Acaphylla theae* (Watt), *Calacarus carinatus* (Green)	*Amblysieus herbicolus* (Chant), *Lestodiplosis* sp., *Scolothrips rhagebianus* Priesner
Bunch caterpillar *Andraca bipunctata* Walker	*Cylindromya* sp.(Techinid fly), *Cantheconidea furcellata*(Wolff), *Pencilium* and *Rhizopus* (Fungi), Bacteria, Dipterous fly
Red slug caterpillar *Eturesia magnifica* Butler	*Apanteles* sp., *Exorista heterusiae*, Bacteria
Looper caterpillar *Buzura suppressaria* Guenee	*Sarcophagus* sp., *Apanteles taprobanae* Cameron, *Aphanogmus monilae*, Bacteria
Flush worm *Cydia leucostoma* (Meyrrick)	*Apanteles aristaeus* Nixon, *Calleida nilgirensis* Straneo, *Asympiesella indis*
Leaf roller *Caloptilia theivora* (Walsingham)	*Sympiesis dolichogaster* Ashmead, *Mestocharella javensis*

Pest	Natural enemy
Tea tortrix *Homona coffearia* Nietner	*Phytodietus spinipes* (Cameron), *Palexorista solennis* (Walker)
Cut worm *Spodoptera litura* (Fabricius)	*Peribaea orbata* (Wiedemann)
Aphid *Toxoptera aurantii,* (Boyer de Fons.)	*Syrphus balteatus, Asarcina aegrota, Scymnus pyrocheilus* (Mulsant), *Verania vincta, Cryptogonus bimaculata, Cryptoonus or Cryptogonus biculus* (Gyllenhal), *Coccinella repanda, Coccinella septempunctata, Lemnia bissellata* (Mulsant), *Pseudaspidimerus circumflexus* (Mots.), *Aphelinus* sp., *Menochilus sexmaculatus* (Fab.), *Jaurevia pubescens* (Fab.), *Aphidius colemani* Vierick, *Lipolexis scutellaris* Mackauer, *Episyrphus balteatus* De Geer, *Paragus tibialis* (Fallen), *Leis dimidiata* var *quindecimmaculata, Trioxys indicus, Chrysoperla carnea* Stephens
Jassid *Empoasca flavescens* Fabr.	*Cephalosporium* sp. (Fungus), Drynid wasp,
Helopeltis *Helopeltis theivora* Waterhouse	*Oxyopes* sp., *Melamphus* sp., *Preying mantis, Chrysoperla carnea* Stephens, Reduviid bug, Spider, Mermethrid Nematode, Ichneumonoids
Scale insects *Eriochiton theae* Green	*Coccinella septempunctata, Spalgis epius, Blastothrix* sp.
Chrysomphalus aonidium Ashmead	*Chilocorus circumdatus, C. nigritus, Pharoscymnus horni, Comperiella bifasciata, Prospeltella* sp., *Aphytis chrysomphali, Aspidiotiphagus citrinus*
Fiorinia theae (Green)	*Jauravia quadrinotata, Scuymnus* sp., *Aphytis theae, Prospeltella* sp., *Aspidiotiphaus* sp.
Aonidiella aurantii (Maskell)	*Aphytis chrysomphali, C. bifasciata*
Saissetia formicarii (Green)	*Coccophagus acanthosceles*

Pest	Natural enemy
Saissetia coffea Walker	*Encyrtus infelix* Embleton , *Coccophagus cowperi* Girault
Shoot mealy bug *Nipaecoccus viridis* (Newstead)	*Aprostocetus purpureus* (Cameron), *Cacoxenus perspicax* (Knab).
Cockchafer grub *Holotrichia* sp.	*Gliocladium roseum* (Fungus).
Termite	*Metarrhizium anisopliae* (nematode).

In a recent study on the presence of functional groups of pests in Makaibari tea estate, the oldest and one of the most respected organic tea estates in Darjeeling hills, it was observed that the largest fraction occupied by the 'herbivores other than pest', and 'pollinators' the smallest fraction. The parasitoids were almost restricted. It was further reported that leaf suckers dominate explicitly in the plantation [12]. It was also observed that seven different groups of food specialist among the tea pests and leaf suckers dominate explicitly in the Makaibari (Fig. 8.1). In another study, greater number of ground level arthropod community (coleopterans) recorded from the organic tea plots (hill)

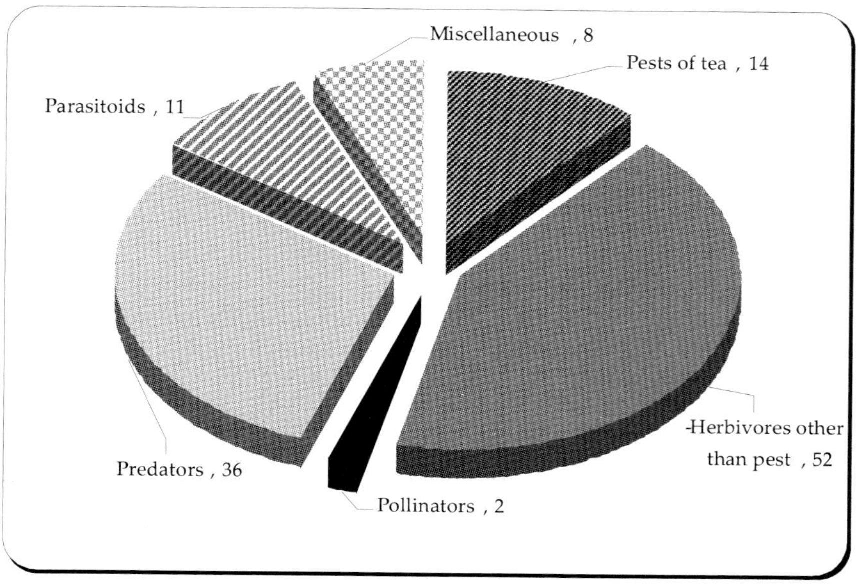

Fig. 8.1. Composition of tropic guilds in the Makaibari tea estate. Drawn with data provided in Wirsig (1999) (Ref. 13).

than conventional (plains) plots (Fig. 8.2). Faunal diversity study at ground level gave the clue that soil of the organic plantation was healthier than that of the conventional tea plots.

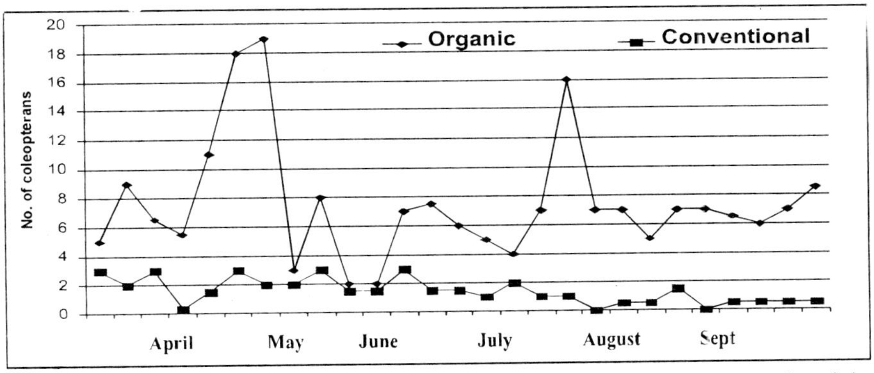

Fig. 8.2. Weekly occurrence of coleopterans (root, borer, cockchafer, white grub etc) in organic and conventional tea plots. Drawn with the data provided in Mukhopadhaya and coworkers [14].

A list of counter-measures against infestation of disease and other pests that are currently being utilized are presented in Table 8.4.

Table 8.4. Counter-measures used against infestation by diseases or pests.

Pest/ disease	Biological counter-measures
Exobasidium vexans (blister blight); Endemic to Southeast Asia, does not occur in East Africa; *Poria hypolateritia* (Red root rot)	Copper preparations permitted in emergencies (max. 3 kg pure copper ha^{-1}); preventive measures (micro-climate, hygiene precautions etc are important;
Meloidogyne ssp. (Nematodes)	Tearing out and burning of infested tea bushes; remove infested tea bushes, and remove and replace large amounts of the soil; prevention e.g. using plant-bags in the seed bed; use shading tree *Indigoferra teismanii* as a trap plant; sow *Tripsacum laxum* (Guatemala grass) before starting a new plantation;
Helopeltis sp. (Tea Bugs e.g. Tea mosquito bug)	In emergencies, neem extracts; encourage useful insects such as ladybirds; introduce *Bacillus thuringiensis*; in severe cases - pruning; always begin the harvest in non-infested sectors;

Pest/ disease	Biological counter-measures
Oligonychus coffeae (Red spider mite)	In emergencies, neem extract, tobacco extract; sufficient shade will suppress development; till weeds early enough before the main harvest begins;
Homona coffearia (Tea roller), an insect that can cause problems in India, Japan, Malaysia and particularly in Sri Lanka.	Plant a diversity of shading tree to attract e.g. parasitical wasps; another natural antagonist is the *Macrocentrus hormonae* parasite;
Andraca bipunctata (Bunch caterpillar) *Biston surpressaria* (Looper caterpillar) *Etrusia magnifica* (Red slug caterpillar)	Counter-act with light traps; collect the caterpillars from the ground, tea bushes and shading trees (in all stages of development); apply trap bands to the shading trees;
Brevipalpus phoenicis (Scarlet Mite) *Calacarus carinatus* (Purple Mite)	Suppress growth with green fertiliser plants and shading trees; apply lime and soda washing in emergencies or after pruning;
Taeniothrips setiventris (Common Thrips; mostly in Darjeeling) *Scirtothrips dorsalis* (Assam-Thrips; mostly in Assam and Dooars)	Green manure plants and shading trees; lime and soda washing; disturbing the soil around the tea bush stem during the cold months will destroy the pupae.
Moss (e.g. in Darjeeling during the winter months)	Wash the tea stems with lime and soda

- According to the European Regulation for Organic Agriculture (EEC) 2092/91 the use of copper preparations for plant protection (e.g. Bordeaux mixture) was allowed for a transitional period up to 31st of March 2002. In case copper preparations have to be applied it is recommended to use preparations which contain less copper and therefore to reduce the accumulation of copper in soils (e.g. tribasic copper sulphate, copper hydroxide).
- According to the European Regulation for Organic Agriculture (EEC) 2092/91 the application of Neem preparations is restricted and only allowed for the production of seed and seedlings. This regulation is discussed controversial. Up-date information is available from certification body.
- According to the European Regulation for Organic Agriculture (EEC) 2092/91 the use of tobacco extracts was allowed for a transitional period up to 31st of March 2002. Furthermore, application is restricted only for tropical and subtropical crops and shall be applied at the beginning of the vegetation period.

The aphid, *Toxoptera aurantii* is naturally controlled by syrphid, coccinellid and hemerobiid predators and aphidiid parasitoids [15]. In China, an adult of ladybeetle *Coccinella septempunctata,* and an adult lacewing *Chrysopa sinica* Tjeder were found to consume 100-120 and 70-90 aphids respectively per day in the laboratory [16].

Scale insects and mealy bugs are subjected to the attack of several parasitoids. *Coccophagus cowperi* and *Encyrtus infelix* heavily parasitise *Saissetia coffeae*[7]. The black scale *Chrysomphalus ficus* is attacked by six parasitoids and predators (www. bioagrosolutions. com).

The leaf roller, *Caloptilia theivora* is heavily parasitized by the eulophid, *Sympiesis dolichogaster*. The incidence of parasitism varies between 20 and 83 per cent (www.bioagrosolutions.com). The braconid, *Apanteles aristacus* is the principal parasitoid of flushworm (*Cydia leucostoma*). This parasitoid, reported from India and Indonesia, appears to be host specific and prefers third instars caterpillars for ovipositio[7]. Population of flushworm in South India is parasitized by nine species of barconid, two ichneumonids and one eurytomid in addition to a pupal parasitoid belonging to *Ascogaster*[17]. Adults and larvae of a cerabid, *Calleida nilgerensis* feed on late instars caterpillars. In South India it was reported that the beetle consumes more than 140 flushworms during its life span[18].

The control of the tea tortrix moth *Homona cafferia* in Sri Lanka in the mid-1930s by an introduced parasitic wasp *Macrocentrus homonae*, still ranks as one of the best classical examples of a successful biological control programme achieved through the introduction of an exotic species of parasite[19]. *Homona cafferia* is affected by one egg, nine larval and four pupal parasitoids[20]. *Apanteles fabiae* and *A. taprobanae* parasitise the looper caterpillar (*Buzura suppressaria*)[21]. Tachinid (*cylindromya* sp.) is the chief larval parasitoid of the bunch caterpillar (*Andraca bipunctata*).

The birds also have good predaceous ability. In China, for example, the fifth instar larvae of the tea geometrid preyed by *Cyanopica cyana* for one time is more than 60, and the tea aphis by *Paradoxyrnis webbiana* were 386[22].

Pest control using entomopathogens

Entomopathogens are inherently less toxic than conventional chemical pesticides and generally affect only target pests and closely related organisms in contrast to broad spectrum conventional chemical pesticides. It is effective in very small quantities and often degrades quickly, avoiding the connected pollution problem[23]. Pathogens can cause epizootics in pest populations, especially caterpillars. Many adults and nymphs of Jassids are found to be attacked by a whitish growth of a species of the fungus *Cephalosporium* in north east India. The dead insect usually remains attached to the under surface of the leaf [24]. Encouraging results were obtained with introduction of *Bacillus thuringiensis* (Bt) against lepidopterous larvae [25], *Homona magnanima* [26, 27], flushworm (*Caloptilia theivora*) and *Adoxophyes* sp. [26]. Its efficacy was also proved against bunch caterpillars (*Andraca bipunctata*) and looper caterpillar (*Buzura suppressaria*) [28, 29] and *Cydia leucostoma* [30]. It has been observed in a field trial in Darjeeling that the symptoms of the *Bacillus* on the bunch caterpillars started only after 24 hours of spray of 10^4, 10^6 and 10^8 per gram spore load of *B. thuringiensis* var *kurustaki*. The percentage of reduction in the population ranged from 44 to 71 percent and 78 to 93 percent after 48 and 72 hours of spray [26]. Bt is safe for human beings but its crystal protein produces a fatal endotoxin in the gut of the target caterpillar pests. To avoid insects developing resistance, several strains of Bt should be combined. Spore load of Bt is reduced quickly in the field after application. Liposome microlipid encapsulation has been used to increase the persistence of Bt in the field.

Infection of cocoons of bunch caterpillar, *A. bipunctata*, by fungi like *Fusarium* sp. has also been reported[31]. Many fungal species of different genera are known to parasitize nematodes. Of these, species of the genera *Arthrobotrys* and *Dactylaria* sp. were recorded in Tocklai, Assam, soils[32]. The fungus, *Verticillium catenulatum* (Kamyshko ex Barron and Onions) W. Gams was isolated from egg masses of root-knot nematodes[24]. A fungus *Paecilomyces tenuipes* (Peck) Samson was recorded in tea estates of Darjeeling parasitising pupal stage of various Psychid pests (Basket worm) of tea during early June to late October[33].

Beauveria bassiana, an entomopathogenic fungus isolated from membracid bug, a shade tree pest, was multiplied and tested against *Helopeltis*. Trials conducted by TRA in Assam and Darjeeling provided 27 percent and 49 percent control respectively by spraying 20 percent suspension for two rounds at 15 days intervals. Coccoidea is the largest group attacked by diverse fungi species of *Aschersonia, Septobasidium, Fusarium* and *Podonectria* [24]. A local strain of *B. bassiana* has been found effective against the shot hole borer in South India. This strain is now being marketed as a wettable powder formulation for the control of shot hole borer [34]. High humidity and cool climate helps in achieving good efficacy of the entomopathogen. *B. bassiana* is also found to infect tea roller, *Paecilomyces* sp. infecting tea geometrid, *Cephalosporium lecanii* infecting many scales, *Cordyceps militaris* infecting the flattened eucleid (*Thosea sinensis*) in China [22]. *B. bassiana* is also used in China to control *Adoxophyes orana* Fischer von Rosierstammand, *Homona coffearia* Nietner and *Myllocerinus aurolineatus* Voss, etc. The control results ranged from 80 to 85 percent [25]. In Sri Lanka, *B. Bassiana* and *Metarrhizum anisoplae* are used against live wood tea termites, *Glyptotermes dilatatus* and *Postelectrotermes militaris* [35, 36].

Box 8.2

Biological Plant Protection in Organic Plantation

In organic plantations, preventive measures have absolute priority. The whole point is not how to eradicate a pest or disease, but how to prevent them appearing. When a plant becomes diseased or is overcome with pests, the planters needs to ask himself how it is that the plant is too weak to thrive in the actual ecosystem without significant damage occurring. The central question therefore remains what the underlying causes are. Possible causes include:

- **The variety used is unsuited to the site**
  It is advisable in such cases to try out different varieties on-site. Yet in addition to its suitability to the conditions on-site, a variety must also meet market demands (quality requirements).
- **The cultivation system is unsuitable**
  This can have many causes. Soil-borne diseases (root-rot, nematodes) are often the cause behind a poor crop rotation, or failing secondary vegetation. Certain fungus types thrive when the plants are set too close together, or in agro forestry systems that are too poorly ventilated. Pests can appear in large numbers when there are no habitats for their natural enemies to grow. Existing hedges or trees on the farm or near to the crops can offer pests hibernation possibilities or hosts during certain of their life-cycle stages.
- **Reduced soil fertility**
  Healthy growth can only come from healthy soil. Poor soil fertility, compression of the soil, water-logging or a too high salinity can all be stress factors for the crops. Stress in plants always results in a higher susceptibility to diseases and pests.

A precise analysis of the actual situation and an alleviation of the causes will lead in both the middle and long term to the most success. This is also because treating the causes will keep the farm largely independent of the use of biological plant protection materials, and thus cut running costs.

Above a certain limit, the appearance of pests or disease in plantation crops calls for measures to protect the harvest. The products for pest and disease control allowed in organic agriculture are listed in the regulation for organic agriculture (EEC) No. 2092/91 as well as the IFOAM Basic Standards. It needs to be stressed, that plant protection products, including those allowed on organic plantations, can only have a limited effect and are not capable of combating the actual root causes.

Source: http://www.naturland.de

Certain commercial formulations of *Verticillium lecanii* (BIO-CATCH V), *Paecilomyces fumosoresus* (PRIORITY), *Metarhizium anisopliae* (BIO-CATCH M) and *Beauveria bassiana* (BIO-POWDER) were evaluated in South India against hairy caterpillar (*Eupterote sp*) and found effective in the laboratory [37]. It has also been reported that application of *Paecilomyces fumosoresus, V. lecanii* and *Hirsutella thompsoni* @ 3500 g formulation per ha significantly reduce the population density of red spider mite in the field [38]. The efficacy of these formulations against the pink mite of tea was also studied and the results were similar to that of above said study [39] Pathogenecity of *P. fumosoresus* on leaf rollers and aphids infesting tea in the lab and field were also found encouraging.

In controlling mites, good efficacy of spraying of wettable powder formulation of *V. lecanii* @ 1500 g/ha during evening hours, when the relative humidity is higher, was found in South India. *Paecilomyces tenuipes* parasitizes pupal stage of various psychid pests (Basket worm) of tea during early June to late October in Darjeeling[33]. Coccoidea is the largest group attacked by diverse fungi species of *Aschersonia, Septobasidium, Fusarium* and *Podonectria.*

In principle, neem extract, *Bacillus thruringiensis* etc are not permitted to use prophylactically in organic farming otherwise pests become resistant. Further measures involving copper preparations must be used sparingly and that too after getting approval of the cetification body.

Different families of viruses cause disease in insects. The major limitation for their use is the concern for human safety. High mortalities of the tortricids, *Adoxophyes* sp. (50 - 88 percent) and *Homona magnanima* (50 - 79 percent) were reported when mixture of two granulosis viruses (GV) isolated from *Adoxophyes orana fasciata,* summer fruit tortrix and *H. magnanima* were applied in the tea field of Japan. BCGV-01 was found promising GV product in Japan. The efficacy of the virus application appeared in the succeeding two generations by suppressing larval density[40]. It was also reported that when GV are sprayed in spring, they are able to control effectively the above mentioned two species of tortrix moths until autumn, but the viruses do not persist in winter[41]. GV is also used for control of nettle grub in Sri Lanka. Nuclear polyhedrosis virus (NPV) is used for control of looper caterpillar in Sri Lanka and China [22]. In field-scale trials, most of the 4th and 5th instars caterpillars were killed in 7-9 days and pest population was reduced from 42 pupae to only 0.03 per plant. NPV,

GV, and CPV (cytoplasmic polyhedrosis virus) of the Retroviridae affecting 18 species of tea pests in China. Some universities, research institutes and large tea farms in China have produced NPV and GV preparations by the live worm bodies, and utilized them to control *Euproctis pseudoconspersa* Strand, *Ectropis oblique* Prout, *Buzura suppressaria* Guenee, *Adoxophyes orana* Fischer von Rosierstamm, and *Andraca bipunctata* Walker, etc. The results showed that the latency of the pathogens was from three to four days, and control effects ranged from 80 to 95 percent [25, 42].

Behavioral control

The use of pheromone for monitoring of tea leaf roller (*Caloptilia theivora*) is indeed very well known. Mass trapping with sex pheromones considerably reduced infestation by *Adoxophyes* both in Japan and Taiwan (Fig. 8.3). Female sex attractants of the flushworm (*Cydia leucostoma*) have also been identified recently. Studies on the sex pheromones of tea mosquito bug in South India (Fig. 8.4) revealed that males were found to be very sensitive and orientation towards pheromone source was optimum. More males were attracted to a few mixtures of acetate and alcohol. Attraction increased with dose and decreased quickly after few days due to volatilization of the components.

Fig. 8.3. Synthetic sex pheromone trap is used to kill adult male tea tortrixes in Taiwan. (Source: http://www.coa.gov.tw/external/teais/)

Fig. 8.4. Sex pheromone study in south India (Source: www. upasitearesearch.

Kairomones from the partly dried stems of *Montanoa bipinnatifida* are found to attract shot hole borer beetles. These stems can be used as a trap to control these beetles in tea fields [34].

Harnessing of volatile synomones

The role of chemical communication in several tri-trophic systems, involving the host plant, the insect feeder (the herbivore) and its complement of natural enemies (predators and parasitoids), is being increasingly recognized as of crucial significance for successful biological control. The host plant emission of specific blends of volatile synanomes, triggered as a result of the feeding activity of the insect pest on the host plant tissue, are vigorously exploited by the relevant natural enemies (predators/parasitoids) to help locate their host insect[43, 44]. These volatiles include a group of C_6 alcohols, aldehydes and their esters [45]. Some volatile components emitted from aphid-damaged tea shoots, as well as to the tea shoot-aphid complex were recently identified in China. These were (*Z*)-3-hexen-l-ol, benzaldehyde, (*E*)-2-hexenal, (*Z*)-3-hexenyl acetate, ocimene, linalool, geraniol, indole, and (*E*)-2-hexenoic acid[46]. Some attempts have been made to use synomones and kairomones in the manipulation of natural enemies to suppress pests[47, 48]. The harnessing of the synthetic analogues of the relevant volatile synomones, hold good promise to help strengthen biological control activity in managing specific tea pests in future.

Use of light traps

Fluorescent light traps are useful in attracting the moths of caterpillars and other insects. At the time of emergence of moths and beetles in the tea fields these traps can be set up, and the attracted moths and beetles can be killed mechanically or by using insecticides. Yellow pan water traps have been found useful to traps thrips and aphids. Yellow sticky traps (Fig. 8.5) have also been found to reduce the thrips attack by trapping the adult thrips[34]. The number of traps per hectare may vary from 70-100 depending on the planting density.

Fig. 8.5. Yellow stick boards are used to kill smaller tea pests in Japan.

Use of trap crops

These crops having specific chemical substances that are toxic to the target pests initially attract the pests to them, thus effectively serving as diversionary hosts, and thereafter kill the trapped pests. In Sri Lanka, *Gliricidia sepium* has been used to trap swarming live wood eating termites[49]. Other trap crops such as *Targets* sp., *Arachis pintoi, Eragrostis curvula, Tithonia diversifolia, Wedeliya trilobata* and *Vetiveria zizanoides* are used to control parasitic nematodes[50].

Use of botanicals

Certain products derived from plants are used for tea pest control. Over 50 years, approximately 2400 plants are reported [51] to possess pest control properties. Some of these plants species possess the characteristics required for an ideal botanical insecticide and are therefore more promosing for use in organic pest control programmes. These have many merits viz. biodegradable, target specific and can be locally produced. They suppress rather than eliminate a pest problem. However, they should be effective at a rate of maximum 3-5 percent plant materials based on dry weight [52].

Botanicals in major use

Neem

Neem (*Azadirachta indica*) has emerged as the single most important source of botanical insecticides [53]. Neem leaves and seed contain azadirachtin, miliantriol etc have been found effective against pink and purple mites, and caterpillar pests such as flushworms and leaf rollers. These constituents are more in seeds than in the other parts of the plant.

Method of preparation of neem seed kernel extracts (NSKE)

- Collect 25 kg of shade dried neem seeds and crush the same to coarser pieces.
- The crushed seeds are tied in a muslin or cotton cloth and immersed overnight in a container in 50 litres of water.
- Extract is then completely separated by squeezing the cloth containing the crushed seeds.
- The cloth containing the crushed seeds is dipped again in 50 litres of water so that the whitish neem seed kernel extract is completely separated from the seeds.
- Mix up 100 litres of concentrated NSKE solution with 400 litres of good quality water. The spray solution now at 5 percent concentration can be used as a spray to repel the insect pests. Depending upon the prevailing conditions of the field, the concentration of the spray can be increased upto 8 percent.
- The spray should wet the whole plant. Even the ventral side of the leaves has to be smeared with the spray.
- The extract should be used within a day or two after making

the formulation. This would prevent degradation of the active ingredients.

- Spray can be done at least two-three times at an interval of 10-15 days.

Method of preparation of neem decoction

20 gram kernel is required for preparing one litre of decoction. This can be obtained from 30 gram dried seed. The crushed kernels may be tied in a cotton cloth and kept in water for about 8-10 hours. The bundle is then to be taken out and pressed to get the extract. The process is repeated till the colour is clear. The extract can be sprayed on the canopies of the plants.

Neem leaf decoction

100 g neem leaves can be boiled in 5 litres of water. Use the solution after cooling.

Neem cake

This can be used in controlling nematodes. 50 g of neem cake may be applied into the pits before planting. The solution prepared by soaking 1 kg neem cake in 5 litres of water overnight can be sprayed.

Neem oil

20 ml neem oil and 1 litre water are required for making 1 litre solution. Take the required quantity of neem oil in a vessel and mix well with thin slices of soap till white foam is produced.

Rotenone

The roots of *Derris elliptica,* and related species viz. *Tephrosia, Lonchocarpus* and *Mundulea* under leguminosae contains 4-11 percent rotenone. Rotenone first appears when the roots are about 6 weeks old and showing suberizatization, roots about 5 mm in diameter are considered to have the highest content [54]. The roots can be used either by drying and powdering or by mixing with water. The spray liquid should contain 0.002-0.004 percent active ingredient. This is useful in controlling sucking pests, caterpillars and some beetles [52]. It acts as a contact and stomach poison [53]. All *Derris* formulations should be stored away from air and bright sunlight. It is extremely poisonous to fish.

Sabadilla

This represents the dried ripe seeds of *Schoenocaulon officinale* (also known as *Sabadilla officinarum* Brant, *Asagraea officinalis* Lindl. and *Veratrum sabadilla* (Ritz.). *Sabadilla* serves as both contact and stomach poison against a wide variety of insects. Crude *Sabadilla* exhibits a low order of mammalian toxicity. The ground seeds containing approximately 0.3 percent alkaloids often caused intense irritation of skin and mucosa and sternutatory (sneezing) effect in human.

Tobacco tea

Cut 400 g of tobacco into small pieces and soak for one day in 4 litres of water. Then apply heat slightly and press the tobacco pieces. Filter the solution to collect the clean liquid and add soft soap @ 1 ml/litre to this liquid. Stir thoroughly and then the decoction may be diluted 5 to 6 times with water and could be sprayed against thrips, scale insects, mealy bugs, aphids, beetles, caterpillars etc.

Pyrethrum

It is extracted from the dried flower heads of *Chrysanthemum cinerariaefolium*. Pyrethrum or natural pyrethroid is a mixture of six different substances and do not have much persistence. They therefore degraded fast and the insecticidal property diminishes. The alkaloid stachydrine is the main active principle. It acts as a contact insecticide and extremely useful in combating outbreaks of pests that multiply rapidly (e.g. aphids). It can be effectively used in controlling thrips, hoppers, scale insects, mealy bugs, ants etc.

Botanicals in limited use

Quassia

Aqueous extract of wood and bark of the small tree *Quassia amara* could be used against insects. Two insecticidal amaroids called quassin and neoquassin have been isolated from *Quassia* and its botanical relatives [55]. These are very effective against aphids. Both contact and stomach poison were observed. Insecticidal extracts were freshly prepared near the point of use by leaching *Quassia* chips with water. For example, 3 kg of the wood was steeped for several hours in 30 litres of cold water and the mixture was boiled for 1 hr and stained. Soft soup (250 g in 10 litres of hot water) was added and the whole was diluted to 100 litres to form 3 percent *Quassia*. *Quassia* chips normally contain less than 0.2 percent of the insecticidal compounds.

Annona

The dried seeds of many species of *Annona* viz. *A. squamosa, A. reticulata, A. muricata* have found to be useful as insecticides. Leaves, roots and bark also exhibit insecticidal property [55]. The leaves and seeds of this plant contain alkaloids viz. lanolin and anonaine having insecticidal property. It acts as a contact and stomach poison. The seeds are to be dried, powdered and made into a solution by mixing with water for application. It is effective against sucking pests, stem borer, scales, hoppers etc.[52].

Botanicals in local use

The extracts (1 percent) derived from the aerial parts (leaves and stem) of *Artimisia vulgaris, Urtica dioica, Polygonum runcinatum, Eupatorium glandulosum* which are growing profusely in the tea plantations of Darjeeling hills are used for their repellant and antifeedant action against some leaf-eating pests [10]. Cut the leaves and stems into small pieces. Take 1 kg of the same in a bucket; pour 10 litres of water and keep for 16-24 hours till fermentation starts. Then filter through muslin cloth or suitable sieves and get the clear liquid. The liquid may be sprayed against pest infestation. Loopings of *Artimisia vulgaris* placed along with the mulch discourage the laying of cockchafer eggs in tea field. Products from plants viz. *Lantana camara, Bidens pilosa, Calotropis gigantia, Ocimum basilicum* and *Vitex negundo* reported to have pesticidal properties [34]. Decoction of *Mimosa pudica* has been reported to possess insecticidal activity [56]. The spray of lemon grass (*Cymbopogon citrotus, C. nardus* and *C. flexous*), citronella (*Cymbophogon winrians*), palmarosa (*C. martini*) (2 percent) and other vegetative oil had good control over some tea pests [57, 58]. Ghosh Hajra [53] listed out some preparations of common botanicals having insecticidal activities. Organic growers like Kiribanda in Sri Lanka spray pounded Khomba (Margosa) seeds mixed with water on the crop. One kilogramme of seed powder is mixed in 20 litres of water. They also rely on birds and pest-eating insects (Source: http://forests.org).

Oilcakes such as those obtained from mustard (*Brassica canipestrei* L.), Sesame (*Sesamum indicum* L.), castor (*Ricinus communis* L.) etc are known to have nematicidal properties[52]. Oil cakes from Undi (*Calophyllum inophyllum* L.) and sal (*Shorea robusta*) were effective against hatching of *Meloidogyne incognita* eggs on tomato seedlings.

Terpenes (mono and sesqueterpenes) and lactones are most promising among the nematicides of plant origin.

Besides oil cakes, the other plant materials, as soil amendments, have proved effective for reducing nematode attack. However, the effective dosage is quite high. The reduction of population of *Meloidogyne incognita* due to soil amendment with neem, *Annona* and *Chrysanthemum* etc. was observed. However, the problem associated with much-held botanicals and bio-pesticides is their varying degree of success in field situation [59].

Cultural control

Probably this is the most economical and widely applicable method of pest control. Certain routine cultural operations such as plucking, pruning, shade regulation and weed control are manipulated to reduce the incidence of pests. Tea mosquito bug laid large number of eggs on the broken ends of plucked shoots. Intensive removal of stalks during plucking is recommended to reduce the incidence of tea mosquito bug. Leaf folding caterpillars can be manually removed while harvesting. A few rounds of black plucking reduce the incidence of thrips and *Helopeltis*. Unshaded areas are more prone to the attack of thrips and mites. The heavily shaded areas will suffer from the attack of *Helopeltis* and blister blight. Proper shade management will help to prevent the excessive build up of thrips, mites and also reduce the incidence of blister blight disease.

Weeds offer excellent hiding places and serve as alternate hosts of many tea pests. *Mikania micrantha, M. cordata, Bidens biternata, Melastoma malabathericum, Eugenia jambolana, Rosa sinensi, Fasminum scandens, Polygonum chinense,* are the alternate hosts of tea mosquito bug in North East India. Effective weed control assumes greater significance in the management of tea mosquitoes. The population of *Pratylenchus loosi* to a certain extent decrease when castor oil and mahua cake were added to the soil [60].

In South India it has been reported that the spread of primary root diseases such as *Poria hypolateritia* and *Fomes noxius* can be prevented by isolating disease patches making 120 cm (4 feet) deep and 45 cm (1.5 feet) wide trenches surrounding the diseased plants [34]. When such isolation trenches are dug the soil from the trenches should be thrown on the perimeter and not broadcast outside the trenches.

Host plant resistance

Planting of pest/disease resistant varieties is one of the important components of organic tea cultivation. Even low levels of resistance are important since the need for other control methods can be minimized. It has been observed that Chineric varieties are more susceptible to the attack of red spider and scarlet mites while Assam cultivars are favoured by eriophyid mites. Incidence of *Helopeltis* is more on China jats in Assam, the clone TV-1 is found highly susceptible to this pest. Flushworm incidence is more on the clone UPASI-17, considered 'cambod' in nature. Soft wooded tea plants are known to be easily attacked by termites. Clones with high content of alpha spinasterol are susceptible to damage by the shot hole borer. Use of certain Sri Lankan selections viz. TRI 2024, TRI 2025 which are popular in South India should be avoided in shot hole borer-prone areas.

In general, clones are less susceptible to blister blight disease, when compared to seedling tea. Certain South Indian selections viz. SMP-1 and SA-6 are somewhat resistant to blister blight disease [4].

Control of tea diseases

Application of beneficial micro-organisms has been attempted in the control of tea root diseases. *Trichoderma viride* Pers. & Fr. and *T. harzianum* Pers. & Fr. showed inhibitory activity against branch canker (*Poria hypobrunnea*) [61]. In North East India when the propagules of *Trichoderma* were spread on the cut surface of stems after pruning and then inoculated with the pathogen, the rate of infection decreased from 29.2 to 3.3 percent after 12 months and from 51 to 5.8 percent after 24 months. *T. viride* and *T. harzianum* are also effective against Charcoal stump rot (*Ustulina zonata*) and brown root rot (*F. noxius*)[6,62]. *Micrococcus luteus* was found to be highly antagonistic to brown blight (*Glomeralla cingulata*) disease in India. In Japan, Horikawa [63] observed that the conidia of *Pestalotia longiseta* and *P. theae* (caused Grey blight) were destroyed by *Streptomyces roseoporus* after three days of inoculation. Control of black rot (*Corticium theae* and *C. invisum*) by strains of *Bacillus subtilis* reported in North East India (69 and 79 percent respectively reduction over control) [7]. Application of *Pseudomonas* strain, isolated from tea soils of North East India, as a bio-control agent against brown rot disease of tea recorded satisfactory control [64].

Weed management

Weeds have traditionally been seen as the enemy, in many cases not without good reason. They interfere with the production of crops, often causing yield losses if not total crop failure. They have to be combated at all costs otherwise they will dominate. However, complete eradication of weeds does not generally occur in the organic farms. There are also conflicts between the measures to control weeds and other aims of the organic systems. For example, early cultivation of pasture to aid weed management may increase the risk of nitrate leaching from the soil [65], and intensive cultivation to kill large weeds may damage soil structure [66]. A balanced weed population can provide a favourable microclimate, and the activities of the plant roots help to improve soil biological activity and structure. Weeds in general are controlled manually by hand pulling during dry periods and slashing during monsoon. Uprooted and slashed weeds should be retained in the field. Weeds can also be managed by different eco-friendly practices.

References

1. Chen, Z. M. and Chen, X. F. 1989. An analysis of world tea pest fauna. *J. Tea Sci.* **9**(1): 73-78.
2. Banerjee, B. 1976. Pesticides and pesticide residues in tea. *Two and Bud* **23**(2): 35–42.
3. Chen, Z. M. and Chen, X. F. 1990. *The Diagnosis of Tea Diseases and Their Control* (In Chinese). Shanghai Scientific and Technical Publishers, Shanghai, Republic of China, p. 275.
4. Muraleedharan, N, 2005. Sustainable cultivation of tea. *Planters Chronicle* **101**(5): 5-17.
5. Ghosh Hajra, N. and Kabir, S. E. 1993. Organic farming – an overview. *Ind. Rev. Life Sci.* **13**: 121-131.
6. Barbora, B. C. 1995. Biocontrol of pests and diseases of tea in N. E. India. *Proc. Intl.Conf. Tea Quality & Human Health Symposium*, Nov. 7-10, 1995. Shanghai, China. 357-362.
7. Agnihothrudu, V. 1999. Potential of using bio-control agents in tea. In: Global *Advances in Tea Science* (ed. N. K. Jain). Aravali Books International (P) Ltd., New Delhi, 675-692.
8. Muraleedharan, N. and Chen, Z. M. 1997. Pest and diseases of tea and their management. *J. Plantation Crops* **25**(1): 15-43.
9. Zeiss, M. R. and Braber, K. D. 2001. *Tea –Integrated Pest Management Ecological Guide*. (Translated by T. T. Nam) CIDSE/Vietnam, p 282.
10. Ghosh Hajra, N. 2001. *Tea Cultivation: Comprehensive Treatise*. International Book Distributing Co., Lucknow, India, p. 518.

11. Babu, A., Selvasundaram, R., Muraleedharan, N. and Sasidhar, R. 2004. A new predator of red spider mites. *UPASI Tea Res. Foundation Newslet.* **14**(1): 1.

12. Hsiao, Suh-Neu 2001. Effect of green lacewig, *Mallada basalis* (Walker) on the spider mites, *Oligonychus coffeae* Nietner and *Tetranychus kanazawai* Kishida in tea plantation. *Proc. Intl. Conf. on O-CHA (tea) Culture Science*, October 5-8, 2001. Shizuoka, Japan, 86-89.

13. Wirsig, A. 1999. Food web and community structure of arthropods in organic and conventional tea gardens of Darjeeling. North East India. *Ph.D Thesis*, University of Hohenheim, Germany, p. 71.

14. Mukhopadhyay, A., Sherpa, P. W. and Pradhan, B. 2003. Diversity of ground arthropod community at organic and chemically intensive tea plantation of Darjeeling terai. *J. Environ. Biol.* **24** (4): 471-476.

15. Radhakrishnan, B. 1989. Studies on the aphid, *Toxoptera aurantii* (Boyer de Fonscolombe) (Hemiptera: Aphididae) and its natural enemies in southern Indian tea plantations. *Ph.D Thesis*. Bharathiar University, Coimbatore, India, p. 154

16. Han B. Y. 2002. Population dynamics of tea aphid, *Toxoptera aurantii* and its natural enemies in tea garden. *J. Tea Sci.* **22**(2): 85-92.

17. Muraleedharan, N., Radhakrishnan, B. and Selvasundaram, R. 1988. Natural enemies in certain tea pests in Southern India. *Insect Science and its Application* **9**(5): 647-654.

18. Muraleedharan, N., Radhakrishnan, B. and Selvasundaram, R. 1991. Observations on the life history and feeding rates of *Calleida nilgerensis* Straneo (Coleoptera: Cerabidae) a predator of flushworm of tea. *J. Ecobiol.* **3**(3): 189-192.

19. Gadd, C. H. 1941. The control of tea tortrix by its parasite, *Macrocentrus homonae*. *Tea Q.* **14**: 93-97.

20. Muraleedharan, N, 1993. Ecology and management of tea pests. In *Tea Culture Processing and Marketing* (eds. M. J. Mulky and V. S. Sharma), Oxford & IBH Publishing Co. Pvt. Ltd., New Delhi, 129 - 144.

21. Dathanarayana, W. and Kathiravetpillai, A. 1969. The bionomics of tea looper (*B. suppressaria*) *Tea Q.* **40**(2/3): 71-83.

22. Hangu, Z. 1987. Natural enemies' resources of tea pest and their ecological control on pests. *Proc. Intl. Symp. Tea Quality and Human Health.* Nov. 4-9, 1987, Shanghai, China, 150-153.

23. Ramarethinam, S. 2004. Role of bio-pesticides in tea cultivation. *Intl. J. Tea Sci.* **3**(3&4): 111-116.

24. Barua, G. C. S. 1983. Fungi in biological control of tea pest and disease

in north east India. *Two and a Bud* **30**(1/2): 5-7.

25. Han B. Y., Dong, W. X. and Cui, L. 2004. Microbial pesticides and insect semiochemicals or control of tea pests in China. *Intl. J. Tea Sci.* **3**(3&4): 117-128

26. Kariya, A. 1977. Control of tea pests with *Bacillus thruringiensis. JARQ* **31**: 213-217.

27. Takatsuka, J. and Kunimi, Y. 2000. Intestinal bacteria affect growth of *Bacillus thuringiensis* in larvae of the oriental tea tortrix, *Homona magnanima* Diakonoff (Lepidoptera: Tortricidae). *J. Invertebrate Pathology* **76**(3): 222-226.

28. Reddy, K. P. S., Chaudhuri, T. C. and Ghosh Hajra, N. 1990. Infectivity of *Bacillus thuringiensis* var *kurustaki* against bunch caterpillar. *Two and a Bud* **37**: 15-16.

29. Borthakur, M. C. 1986. Role of entomopathogenic bacteria for the control of tea looper caterpillar. *Two and a Bud* **33**: 1-3.

30. Bisen, J. S. and Ghosh Hajra, N. 1997. Testing of delphin with some commonly used insecticides to control flushworm in young tea of Darjeeling. *Pestology* **30**(3): 12-14.

31. Banerjee, B. 1979. A factor analysis of the population fluctuations in Andraca bipunctata Walker (Lep: Bombycidae) *Bull. Entomol. Res.* **69**: 195-201.

32. Agnihothrudu, V. 1961. A comparison of some techniques for the isolation of fungi from tea soils. *Mycopath. et Mycol. Applicata.* **16**: 235-242.

33. Debnath, S. 1986. A new record of an entomogenous fungus *Paecilomyces tenuipes* (Perk) Samson on various Psychid pests of tea in Darjeeling. *Two & A Bud* **33**(1-2): 24-25.

34. Muraleedharan, N, 2004. Strategies for reducing pesticide residues in tea. *Intl. J. Tea Sci.* **3**(3-4): 105-109.

35. Vitarana, Sushila, I. 1988. Feasibility of biological control of the low-country live-wood termite (*Glyptotermes dilatatus*) of tea (*Camellia sinensis*) in Sri Lanka. In: *Proc. Regional Sci. Conf.*, Colombo, Sri Lanka, 19-22 January, 1988. pp. 99-112.

36. Amarasinghe, L. D. and Abeyasinghe, L. S. 1995. Laboratory and field studies with biopesticides and a chemical insecticide against live-wood tea termite *Postelectrotermes militaris* (Desneux). *Proc. Fifteenth Annual Session of the Institute of Biology*, Sri Lanka, 2.

37. Ramarethinam, S., Loganathan, S., Marimuthu, S. and Murugesan, N. V. 2001. Effect of the entomopathogenic fungi against the hairy

caterpillar, Eupterote sp. (Eupterotidae: Lepidoptera) infesting tea. *Pestology* **25**(5): 1-6.

38. Muraleedharan, N, 2001. Evaluation of three entomopathogenic fungi for the control of red spider mite. *Proc. Intl. Conf. on O-CHA (tea) Culture Science*, October 5-8, 2001. Shizuoka, Japan, 82-85.
39. Selvasundaram, R., Sudarmani, D. N. P. and Muraleedharan, N. 2001. Entomopathogenic fungi for pink mite control in tea. *UPASI Tea Res. Foundation Newslet.* **11**(1): 2.
40. Kodomari, S. 1993. Microbial control of tea insect pests in Japan. *Proc. Intl. Symp. Tea Sci. Human Health.* Tea Research Association, January 11-14, 1993, Kolkata, India, 153- 157.
41. Kawai, A. 1997. Prospect for integrated pest management in tea cultivation in Japan. *JARQ* **31**: 213-217.
42. Liu xin, Shangwen, F., Zhangyu, Amin, S. and Zhichang, W. 2004. The development of organic tea in Chinese mainland. *Proc. Intl. Conf. on O-CHA (tea) Culture and Sci.*, Nov. 4-6, 2004 Shizuoka, Japan, pp. 732-735.
43. Norland, D. A., Lewis, W. T. and Altieri, M. A. 1988. In: Novel Aspects of Insect Plant Interactions (eds. P. Barbosa & D. K. Letoumeau). John Wiley & Sons, New York, 65-90.
44. Turlings, T. C. J., Tumlinson, J. H. and Lewis, W. J. 1990. *Science* **250**: 1251-1253.
45. Sivapalan, P. 1999. Pest management in tea. In: *Global Advances in Tea Science* (ed. N. K. Jain). Aravali Books International (P) Ltd., New Delhi, 625-646.
46. Han B. Y. and Chen, Z. 2002. Behavioral and electrophysiological responses of natural enemies to synomones from tea shoots and kairomones from tea aphids, *Toxoptera aurantii*. *J. Chem. Ecol.* **28**(**11**): 2203-2219
47. Xu Ning, Chen, Z., Yin, K., Wang, H. and You, X. 1995. Induction of volatile parasitoid attracting synomone in tea plants by feeding of tea geometrid (*Ectropis obliqua hypulina*) larvae and behavioral response of *Apanteles* sp. to different odours. In: *Proc. Intl. Tea Quality & Human Health Symp.* Sanghai, China, 1995. pp. 381-385.
48. Bakthavatsalam, N. and Singh, S. 1996. L-tryptophan as an ovipositional attractant for *Chrysoperia carned* (Stephens) (Neutroptera: Chrysopidae). *J. Biol. Control* **10**: 21-22.
49. Sivapalan, P., Seneratne, K. A. D. W. and Karunaratne, A. A. C. K. 1977. Observations on the occurrence and behavior of live wood termites (*Glyptotermes dilatatus*) in low-country tea fields. *PANS* **23**(1): 5-8

50. Gnanapragasam, N.C. 1997. Plant derivatives in managing nematodes attacking tea in Sri Lanka. *Intl. J. Nematol.* **7**(1): 111-112.

51. Grainge, M. and Ahmed, S. 1988. *Hand Book of Plants with Pest Control Properties.* John Willey & Sons, New York, p. 47.

52. Radhakrishnan, B. 2005. Indigenous preparations useful for pest and disease management. *Planters' Chronicle* **101**(4): 4-16.

53. Ghosh Hajra, N. 1993. Botanical pest control: An alternative for toxic pesticides. *APSI Monograph,* Academy of Plant Sciences India, Muzaffarnagar, p. 34.

54. Purseglove, J. W. 1968. Leguminosae. In: *Tropical Crops-Dicotyledons.* Longman Group, London, p. 719.

55. Crosby, D. G. 1971. Minor insecticides of plant origin. In: Naturally occurring insecticides. (eds. M. Jacobson and D. G. Croby), Marcel Dekker Inc., New York. P. 573.

56. Upawansa, G. K. 1989. Ancient methods for modern dilemmas. *ILEIA Newslet.* **3**: 9-11.

57. Ghosh Hajra, N., Kabir, S. E. and Bisen J. S. 1994. Botanical pest control-an alternative to minimise residue toxicity in Darjeeling tea. *Proc. Conf. Plant Protection & Human Health,* March 2, 1994, DPA, Darjeeling, India, pp. 58 - 74.

58. Mohapatra, L. N. 2001. Evaluation of some aromatic oils and plant extracts against tea mosquito bug *Helopeltis antonii* Signoret. *Pestology* **24**(5): 22-23.

59. Ghosh Hajra, N. 2002. Biocontrol of insect pests in tea plantation – An Overview. *Pestology* **25**(5): 9-18.

60. Gnanapragasam, N.C. 1991. Influence of soil amendments in reducing pathogenecity to tea by root lesion nematode. *Pratylenchus loosi. Proc. Intl. Tea Sci.* Tea Exptl. Station, August 26-29, 1991, Shizuoka, Japan, pp. 684-688.

61. Barua, K. C., Borthakur, B. K., Dutta, P. and Begum, R. 1989. Bio-control of *Poria* stem disease of tea caused by P. hypobrunnea. *Proc. Natl. Symp. N. E. Hill University,* Medziphema Campus. 28-31.

62. Hazarika, D. K.; Phookan, A. K., Saikia, G. K., Borthakur, B. K. and Sarma, D. 2000. Management of charcoal stump rot of tea with bio-control agents. *J. Plantation Crops* **28**(2): 149-153.

63. Horikawa, T. 1988. Occurrence of tea grey blight and tea shoot blight caused by *Pestalotia longiseta* Spegazzini and control. pp. 181-195. In: *Proc. Intl. Symp. Recent Development in Tea Production.* Taiwan Tea Expt. Stn., Taiwan.

64. Misra, A. K., Dutta, S. and Dileep Kumar, B. S. 2004. Biological control of brown rot disease and enhanced crop production in tea by a fluorescent *Pseudomonas* strain. *Intl. J. Tea Sci.* **3**(3-4): 338.

65. Cussans, G. W. 1992. The implications for weeds and weed control, of practices to reduce nitrate leaching. *Aspects Appl. Biol.* 30, Nitrates and Farming system, 351-358.

66. Colquhoun, J. B. and Bellinder, R. R.1996. Re-evaluating cultivation and its potential role in American vegetable weed control. *Proceedings X[e] Colloque International Sur la Biologie Des Mauvaises Herbe,* Dijon, France, pp. 335-341

Chapter 9

Harvesting and Processing

Harvesting (plucking) is the most important operation and intricately related to pruning and tipping. In tea, the tender apical portions of shoots comprising the terminal bud, the internodes and two or three expanding leaves immediately below it, which together constitute the crop, the 'flush' (Fig. 9.1) is removed at periodic intervals. It has been worked out that harvesting the first two leaves together with the unfurled bud is the best compromise between the conflicting demands of high yield and high quality [1]. The time interval in days between pluckings is known as the 'plucking round'. This has a cardinal role in harvesting maximum crop without impairing the quality. The tea pluckers or pickers carry the plucked leaves in baskets or bags on their backs and at the end the leaves are transported to the factory for processing or manufacturing.

Fig.9.1.Tender shoots – flush.

Plucking provides stimulus and ensures regeneration of shoots if a steady supply of assimilates to growing buds is ensured by the retention of adequate photosynthetic machinery of the bush.

Handling of harvested leaf

Generally, leaf after plucking certain quantity/full basket taken to a collecting-point and weighed and then moved to the factory for processing. Withering in fact starts as soon as the leaf is plucked. The careless handling between the time of plucking and its arrival at the withering troughs is one of main causes of the loss of quality. Pluckers when plucking leaf in the field damage the same by grasping it too tightly in their hand, and bruising it when putting into baskets. Plucked leaves are generally packed either in coir bags or plastic containers. A variety of means of transport are used for transportation of leaf to the factory. While transporting in lorries or tractor-drawn trailers, the bags should not be placed one above the other and to enable more bags to be put on a truck a removable intermediate floor is advisable.

Types of tea and processing

Tea is generally categorized into three different types - black, green and oolong. Black tea is consumed in Western hemisphere, while consumption of green tea is confined mainly to Asia and Middle East. The worldwide black tea production is about 78 percent, whereas green tea constitutes about 20 percent and rest about 2 percent is oolong tea. Oolong is mainly produced in China and Taiwan. Small quantities of green and oolong are also produced in India and Sri Lanka.

Black tea

Manufacturing processes vary from country to country, but there are two basic methods - orthodox and CTC (cut, tear, and curl). All black teas, both orthodox and CTC (Fig. 9.2), go through four basic stages viz. withering, rolling, fermentation (oxidation), and firing.

Fig. 9.2 Liquor of black tea.

Withering

It is the first stage of tea manufacture. During withering harvested tea flush is evenly spread on the withering troughs which are normally 4 to 5 feet wide and 50 to 70 ft long (Fig. 9.3). During the process hot and cold air is blown through to remove the moisture of fresh leaf which is achieved during 16 to 24 hours of period.

Fig. 9.3. Plucked tea leaves on withering trough.

Rolling

This process has two objectives: to reduce the leaves into small pieces and to disrupt the leaf structure, so that the polyphenol oxidase is brought into contact with the flavonols and the fermentation can proceed efficiently.

In case of 'orthodox' type of tea, the withered leaf is charged in the rollers which vary in size (Fig. 9.4). Normally they are 36" and 48" rollers capable of rolling 70 to 120 kg of withered leaf. The pressure is exerted on the leaf in 'open and pressure' sequence. This operation lasts for about 45 minutes. During this process tea leaf is twisted as a result of repeated rolling action and simultaneously leaf tissue ruptures thus forcing out the cell-sap which spreads as thin film on the surface of the leaf. Various enzymes and their substrates come in contact and enzymatic oxidation starts at this stage and imparts liquoring properties to the leaves i.e., briskness, brightness, colour, strength, and quality to the made tea. Moreover, in case of orthodox tea manufacture, this process imparts twist or style to the made tea. The green colour of leaf starts diminishing and brown coppery colour of leaf begins to appear.

Fig. 9.4. Rolling machines inside an orthodox tea factory.

The CTC (cut, tear and curl) machine has two rollers, carrying teeth, which rotate in opposite directions at different speeds. The teeth mesh in with one another, leaving a gap that is variable. Leaves are drawn through the gap and are cut and curled. It is common practice to reduce the size of the leaves fed to a CTC machine by prior passage through a rotorvane. The cut leaves often pass through a second CTC machine. These machines reduce the leaves to very much smaller pieces than those described earlier. This is important for the fine tea required for tea bags (Fig. 9.5). The tea leaves are moved up to the top of the machine

Fig. 9.5. CTC machines in a tea processing factory.

by the inclined moving belt. The cutters are behind the tops of the inclined belt guards. The tea falls between the pair of cutters. The first machine on the right discharges cut leaf on to the feed belt of the second machine.

Fermentation (Oxidation)

It is mainly an enzymatic process during which polyphenols are oxidized with the help of enzyme and oxygen from air. The main object is to complete the chemical change up to the required degree so that the best balance of liquoring properties is obtained to make the tea liquor pleasant to taste. Rolled leaf is put on racks specially made of tiles (Fig. 9.6), cement or aluminium. Oxidation rooms should be kept clean and cool. In Darjeeling, oxidation is carried out at low temperature. During hot season the air is kept humid with the help of humidifiers with less than 20^0c hygrometric difference. Optimum oxidation is adjudged by the subjective assessment of the tea maker. Oxidation period varies from 2 to 3 hrs depending on the type of leaf, degree of wither, temperature, availability of oxygen and oxidizing ability of the tea flush. The time of oxidation is reckoned from the time at which rolling begins. Oxidation is a very sensitive process. Even little delay in adjusting the correct oxidation period is likely to deteriorate flavour appreciably. During oxidation the leaf changes colour and becomes dark coppery. A typical aroma develops at this stage. Oxidation of catechin/ polyphenols by the enzyme polyphenol oxidase is the main reaction of the process that imparts 'tea character' to the black tea.

Fig. 9.6. Rolled leaf on racks of Ambootia.

Firing / Drying

The main objects of drying tea are to stop the enzymatic mellowness of fermented leaf and to obtain the desirable mellowness of resultant tea, and to reduce the moisture of fermented leaf to 2.5 to 3 percent. Thickness of spread, speed of trays and volume of air blown through are regulated meticulously so as to achieve the correct drying. The whole process of drying is completed within 20 to 30 minutes (Fig. 9.7). Darjeeling teas are normally fired at high temperature which is considered essential for squeezing the correct balance of volatile flavoury compounds in high quality teas.

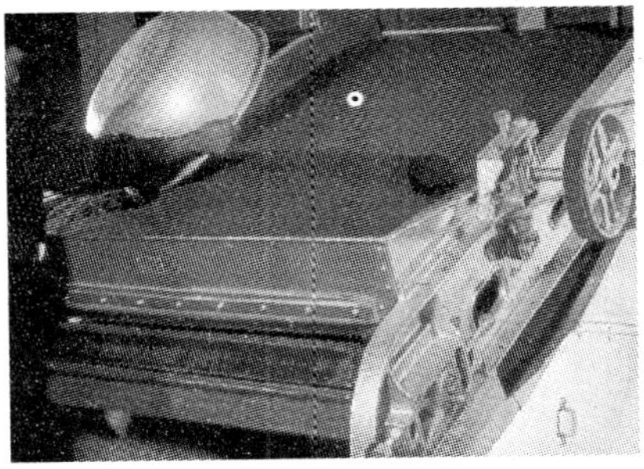

Fig. 9.7. Multi-band Drier.

Sorting and grading

At the end of the manufacturing process, the leaf particles are sorted into different sizes or 'grades'. Bulk tea is sorted on mechanically oscillated sieves. One of the objectives of sorting is to classify the tea as per the size of the particles. The sieve size decreases gradually from top to bottom. Thus, the bigger size teas are retained on the top while the broken leaves fall towards the floor (Fig. 9.8). The modern sorting machines employ different sieves having meshed nos. 8, 10, 12, 16, 24, 30 or more. Teas are passed through electrostatically charged rollers (Fig. 9.9) which attract the fibre and stock preferentially.

Fig. 9.8. Multi-deck sloping type of sorting machine.

Fig. 9.9. Fibre-removing rollers.

Thus, a series of clean teas free of stalk and fibre are produced. The tea passing through the meshes are classified into various grades. In general, there are four grades of tea. They are Whole leaf, Broken leaf, Fannings and Dust. Each of these grades is again classified into various sub-grades. The key grades of black tea are: CTC grades - BP1, BP (Broken Pekoe), PF1, PF (Pekoe Fannings), PD (Pekoe Dust), D1, D; Orthodox grades - TGFOP (Tippy Golden Flowery Orange Pekoe), GFOP, FOP, OP; Brokens - TGBOP (Tippy Golden Broken Orange Pekoe), GBOP, FBOP (Flowery Broken Orange Pekoe); Fannings - BOPF, GOF, Dust - PD, D1 etc.

Green tea

The main green tea-producing countries are China (546,000 tons), followed by Japan (84,000 tons), Indonesia (40,000 tons), and Vietnam (33,000 tons). Green tea is mainly consumed in China, Japan, and South East Asian countries, especially by the Chinese residents, and some countries viz. Morocco, Tunisia, and Libya. The plucked leaves are laid out in the sun or in the warm air of the factory and are allowed to wither until they are slightly limp and soft. The leaves are then heat-treated to destroy any enzymes that would react with the oxygen in the air and cause the leaf to rot. Green teas are therefore known as unoxidised teas. Heat is applied either by pan-frying in a large wok, as in China or by steaming, as in Japan. The fresh green leaves are steamed to kill the oxidizing enzymes allowing the final product to stay green. The steamed tea is then alternately rolled and dried, but allowed to retain 3 percent moisture. Steaming is the common practice in Assam, whereas in Dehra Dun in Uttaranchal and Kangra valley in Himachal Pradesh the leaves are fan fried. In India, rolling is done in a conventional roller unlike the Japanese process in which the rolling is done in a drum. Green tea leaves gives straw coloured or pale greenish-yellow (Fig. 9.10) or brownish-yellow liquor and generally have a somewhat astringent or pungent flavour. The common grades of green tea are: Fine Young Hyson, Young Hyson, Hyson No. 1, Twanky, Fannings or Soumee, and Dust.

Fig. 9.10. Liquor of green tea. Courtesy: www.teahealth.co.uk

Oolong tea

This tea is often called semi-fermented or semi-oxidized tea and is principally produced in the Fujian province of China, and in Taiwan. The time of picking is crucial for the production of quality oolong tea - the leaves should be neither too young nor too mature. The leaves are wilted for four to five hours in direct sunlight or in warm air inside the factory to remove some percent of the water content. The leaves are then rolled and allowed to react with oxygen in the air. The leaves start to give off a smell similar to orchids or peaches. The colour of the leaf gradually changes and the centre of the leaf remains a yellowish green while the edges, where the enzymes have oxidized, become reddish brown. The processed leaf is heated either to pans or ovens to stop oxidation and allowed to retain 3 percent moisture content. Strawberry coloured liquor.

White tea

It is an unfermented tea made from very young tea leaves or buds that are steamed immediately after harvest to inactivate polyphenol oxidase and then dried. Consequently, white teas usually contain higher concentrations of catechins than other teas (http://lpi.oregonstate.edu/f-w02/tea.html). Tiny silver hairs that cover the buds give the tea a white appearance. White teas are the rarest in the world, produced on a very limited scale in China and Sri Lanka. Traditionally plucked only at daybreak in four provinces of North East Fuijan, so that the buds are unopened.

Post-harvest and manufacturing practices

Isolation of manufacturing facility

For estates having both conventional and organic farming activities, manufacturing of organic tea should be carried out in a separate factory to eliminate all chances and possibilities of coming into contact with the conventional tea. The manufacturing of 'organic tea in conversion' and 'certified organic tea' should preferably be done on separate days; utmost care should be taken to properly clean and wash the factory with water under pressure before the manufacture of 'organic tea' and after manufacture of 'organic tea in conversion. Use of additives is not permitted during the process of manufacture of organic tea.

The processing unit of organic tea should possess ISO certification by the concerned accreditation agency.

Storage and packing

Organically produced and manufactured teas should be stored preferably in separate storerooms and warehouses where no fumugants, insecticides or fungicides are used. Vacuum, steam or high pressure water cleaning is permitted. In case of small organic tea producers, it would be worthwhile to have common storage facilities of a smaller capacity if they are interested in holding their tea for better returns. This would also facilitate common inspection by the certifying body.

Packaging material for organic tea must be biodegradable and eco-friendly. Waste generating package material should be avoided. Labeling should clearly indicate the information about organic tea packed. When the full standards requirements are fulfilled, products can be lebelled as "Produce of Organic Agriculture".

Possible contamination source

Proper care must be taken to ensure that the produce does not become contaminated by foreign substances. Possible contamination sources include:

- Substances (e.g. copper, lead from abrasion) emanating from processing machinery that the tea comes into direct contact with;
- Wood protection preparations used to protect wooden crates (e.g. PCP);
- Glues used to make the crates (often containing formaldehyde);
- Glues used in consumer packages often contain contaminants (e.g. PCP);
- Ink used in crates and bags often contain contaminants.

Transportation and shipment

The chests of organic tea should be transported separately and there should not be any chance of it coming into contact with the conventional tea. Before shipment to destination, organic tea chests should be stored in a separate place, away from the conventional tea.

Reference

1. Willson, K. C. 1992. Field operations: 2. In: *Tea: Cultivation to Consumption*. (eds. K. C. Willson and M. N. Clifford). Chapman & Hall, London, 227-226.

Chapter 10

Transitioning to Organic

The trajectory of switchover from conventional to organic is not smooth and in fact obstacles are numerous. The first and foremost among them is definitely the financial and economic costs involved in the process of conversion. The decision to convert all or part of a tea estate requires a high level of commitment on the part of the farm workers and supervisors if it is to succeed. Despite these, the decision to convert carries with it a high element of risk and uncertainty as far as the financial viability of the tea estate is concerned, and this is further compounded by the current lack of detailed information and advice. In the initial years of conversion, a wide range of problems might arise, and these could seriously undermine the confidence of those undertaking the conversion. It is at times like these, that the support of close friends, family, other organic producers, neighbours and others who have an interest of the wellbeing of the respective tea estate becomes vital. Further, before the management system of the tea estate changes, it is vital that the farm workers have to be convinced that the new system is better than the old one. Otherwise, the conversion process really becomes difficult.

The conversion from conventional regime to an organic system primarily requires complete deterrence from application of chemical inputs. Furthermore, the chemical residues left in the soil by the hitherto practiced conventional agricultural techniques should also be neutralized. However, the organic conversion process is not just about replacing a high-input chemical system with a low-input non-chemical alternative. It is, in fact, much more than just abandoning application of chemicals. It entails, in most cases, quite radical and dramatic changes to the farming system as a whole and requires a fundamentally different approach to the management of tea estate.

In the past, it has been observed that the conversion in some cases has usually been done in a very haphazard manner. There are in fact examples where tea planters have simply stopped applying nitrogen and other chemicals and then found that crops fail. The crop has also succumbed to massive pest infestations. However, the situation has now changed markedly. Information is more readily

Box 10.1

Attitude, Prospects and Barriers to Cultivating of Organic Tea

- Tea planters have become more aware of environmental problems (erosion, pesticide residues in tea and several health hazards connected with an intensive system of tea production);
- demand for organic tea has grown constantly as a result of increased consumer awareness of pesticide residues, heavy metals etc in non-organic teas;
- some planters believe organic tea-growing is not a fad anymore but the need of the day for increasing the productivity and sustainability;
- planters who are considering a conversion to organic cultivation at a later date have a fear on potential decrease in yield. In addition, apart from yield declines depending on the condition of the garden, variable cost increased by about 50 percent. Research and development in organic agriculture to improve productivity is relatively scarce. The increased labour requirement is also found as a barrier to adoption. The production patterns tend to be more diversified;
- high cost of organic certification and enormous amount of mandatory documentation involved in the process of inspection and certification, which is too cumbersome to maintain for those small growers, who are illiterate;
- there is scarcity of organic manures at present. Further, these are quite expensive especially when it involves transportation from production site to the plantation site;
- inadequate availability of biofertilizers, biopesticides etc;
- deficiency of potassium and zinc in organic tea fields;
- difficulty in controlling insect pests like *Helopeltis*, mites, disease viz. blister blight etc;
- constraints on access to international markets;
- lack of adequate marketing channels, which prevents an organic tea producer from securing the premium of produce. If there is no assured market, this tea will have to be sold at rates similar to conventional teas;
- planters do not normally obtain any price premium during conversion period. Due to the expected lowering of yields between 30-35 percent and higher production costs, the transition to organic production may lead to enormous financial losses within this period;
- planters are unsure whether the premium will remain at least at the current level. There is still fear that large scale conversion would lead to massive oversupply, a collapse in price and reduced income;
- there was the steady narrowing of price margins between good quality non-organic and organic tea. This is because even non-organic gardens use relatively low levels of pesticides. Moreover, most non-organic tea growers claimed that, as tea was just an infusion, there was not much difference between the taste of non-organic and organic tea;
- lack of access to relevant knowledge information;
- dearth of training facilities and the non-existence of an adequate extension system;
- institutional barriers, such as, scarcity of professional institutions capable of assisting the planters throughout production, post-production and marketing processes.

available nowadays, particularly as a result of the formation of different producer organizations, through certifying agencies which are also providing organic advisory services. Research work specific to organic agriculture is being conducted by different national and international institutions.

Conversion planning

An analysis of the experiences of the tea planters who have converted their tea estates to organic systems indicates that the main problems which are encountered during conversion process include reduction in yield, infestation of pests, unexpected high workloads in peak periods and finanacial difficulties. Planning the conversion entails being aware of the problems which can occur and attempting to avoid them by preparing for them in advance. The preparation of conversion plan is normally done by the proprietor or an advisor in conjunction with the manager and supervisors of the tea estate workers (Kamdaris). The first stage is an assessment of the motives behind the conversion and whether they form a suitable basis for a successful conversion, together with an analysis of the current situation on the estate, including size and layout of the estate, soil texture and nutrient status, manure handling systems, livestock availability in the estate, rainfall, workers and capital availability. In converting to organic garden, labour inputs increase by about 30 per cent per year. During conversion additional labours are required for trenching, compost preparation, weed control, application of bulky and concentrated organic manures and shade regulation. Other costs, such as planting of trees, grasses, *Citronella* etc, application of manure, organic pesticides and bio-fertilizers, depend upon the extent to which integrated farming can be practiced For tea, where chemical fertilizers are applied at a few hundred kg per hectare, compost and cakes are required in tons per hectare, shooting the cost of labour upward. In an industry where the cost of labour can be as high as 60 percent of the total cost of production, a significant increase in labour cost can pose a serious financial risk. The second part involves the development of a target organic system and a plan for the transition between the current, conventional system and the organic endpoint [1]. Technical questions, such as measure for soil improvement, direct weed, insect pests and disease control techniques, appropriate cultivation practices and suitable manure management systems are not very important at this stage of planning process, although they do need to be considered at some stage before the conversion process gets underway.

Livestock is an integral feature of organic framing for cow dung, sheep, goat, poultry and pig manure (see also chapter 7 page 93). If there is sufficient livestock in the garden, adequate supplies of manure are available, and actual declines in variable costs are realized. Capital investment towards purchase of livestock (if not sufficiently available in the estate), improved livestock housing, feeding and manure handling systems as well as facilities for processing, packaging and marketing of processed tea might also be necessary.

The tea estate undergoing conversion cannot even take advantage of the premium prices available for organic products because during this official transition period the tea cannot be sold as 'organic'. As a result, they cannot expect to cover for the initial financial expenses of conversion. Even though sometimes this tea are allowed by the certifying authority to be labeled and sold as 'organic in conversion' tea, it can at best fetch prices slightly higher than conventionally-produced tea. Such prices can hardly compensate for the financial burden of conversion. Moreover, even if such labeling and sale is allowed, significant market development costs may be associated with marketing of such tea [2].

Only when the target system has been developed within an agricultural/ecological context is it really appropriate to consider the financial implications in detail. Obviously, no system could be expected to be adopted which is not also financially viable, but the primary consideration should be a system which works agriculturally and is ecologically sustainable, with as few compromises imposed by economic constraints as possible. In this sense, a satisfactory level of income to ensure the financial survival of the tea business and meet the needs of the tea estate family is a more appropriate goal than profit maximization which may, and often does, conflict with environmental considerations and sound agricultural practice [1].

The simplest way to deal with financial implications is to calculate gross margins on the basis of estimated yields and current prices and premiums as an attempt to represent the situation when the tea estate is finally converted. A more detailed financial assessment would involve fixed costs, capital investment on machinery purchases, additional housing for livestock etc. A further stage could involve the preparation of annual budgets during the conversion period.

A planning process does not only involve the design of a feasible, "target" organic system. It must also set out the steps that need to be taken during the conversion process and the timescale over which the conversion takes place. Probably the most important consideration when developing a timescale for conversion is to start slowly. Experience needs to be gained with new cultivation practices, new techniques and the potential output from the system. Once some experience has been gained, more blocks can be converted. However, it is of course possible to attempt a conversion of the entire tea estate at once, but the risk inherent in this approach are great and it is not advisable unless there are specific circumstances which warrant it.

In organic tea estate, prices and premiums, and yield trends are important for organic tea growers. Theseare also vital for the growers who are considering a switchover from conventional to organic management. Therefore, an idea on the price premiums and yield trends during and after conversion has been described below.

Price of organic tea

The price of tea in general is largely determined by supply and demand. Tea corporations such as Brooke Bond Lipton Ltd., (Unilівer), Lyons, Tetley and Premier Brands have a considerable influence on supply and demand and thus on the price-fixing process. Market concentration is extremely high. 90 percent of Western trade is in the hands of seven transnational companies and 85 percent of world production of tea (all kinds of tea) is sold by transnationals [3].

The price for organic tea strongly follows the market rules and, similar to the price of conventional tea, is closely connected to quality. While on the conventional tea market the price is relatively stable due to a balanced supply and demand situation, the small organic market is characterized by enormous fluctuation. The probable reasons are:

1. A large number of tea estates converted to organic production due to the advice of the international certifying bodies. This resulted in enormous increase in volume of organic tea in recent years. At the same time, demand grew, in line with other product groups, at between 10–20 percent;
2. The second main reason concerns the permanently changing consumer trends and tea variety preferences. Whereas in the late 1990s, green tea was the trendsetter for a healthy lifestyle, today teas, like 'Rooibus' or 'Lapacho' have become more attractive for consumers. After the start of the green tea boom in Europe, countless small farmers in China converted to organic

farming. After a time lag (period of farm consulting and conversion) currently just 20 percent of all Chinese organic green tea can be sold as organic tea. This situation has led to a drop in prices for producers. For example, the premium price for organic tea in 1999 was 300–400 percent (USD 6-8/kg) on top of the price for conventional tea, the premium dropped to 20–30 percent (USD 2/kg) (www.sippo.ch). For most tea producers, this market price is far too low to cover the production costs (USD 6/kg minimum). Comparatively, it is interesting to note that the consumer prices for organic tea have remained quite stable. This means, despite the existing oversupply situation, the international traders can make profits. However, more serious traders generally have long-term contracts with their tea producers and guarantee them stable purchase volumes and in some cases also minimum prices. This has led in the current market situation to enormous price and income differences between farmers producing in the same region, and in some cases to significant differences in consumer prices.

Box 10.2

FAO Tea Composite Price

The Food and Agricultural Organization (FAO) composite price of tea averaged US$ 1.65 per kg in 2004, about 9 percent higher than 2003 (Rs. 45 per = US$ 1). The composite price in March 2005 averaged US$ 1.71 per kg, a 9.6 percent increase over the price reached in February, largely due to firm demand against smaller traded volumes in Mombassa and Kolkata auction markets.

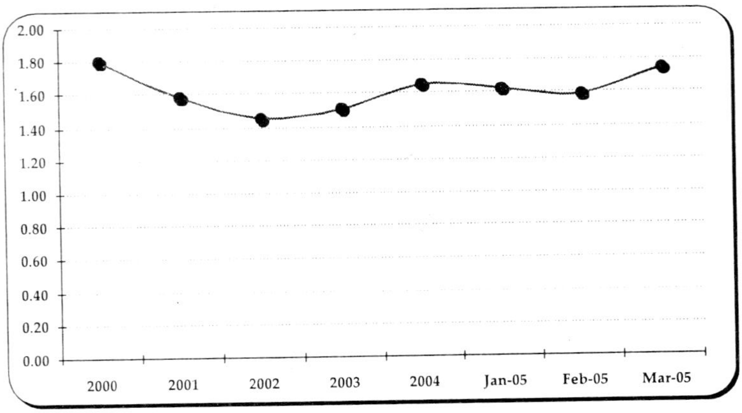

Source: The Food and Agricultural Organization, Rome (www.fao.org)

Organic products are generally more expensive than conventional products. The consumers are willing to pay price premiums for organic products. It is an accepted fact that organic suppliers should point out all premium benefits for their products in a powerful promotional campaign to get optimum price of the product. The premium prices can make organic tea estate economically as well as environmentally attractive; they can only be achieved where a viable marketing strategy has been established. In early 90's the price difference between organic tea and non-organic tea in India however went up 200-300 percent but later has stabilised at around 50-100 percent (Table 9.1). It is not unlikely that due to the expected increase in organic tea production and the existing oversupply situation on the world market, the price of organic tea in India may decline to 30-40 percent in the coming years (www.sippo.ch).

Table 10.1. Increase in price of organic black teas over non-organic ones (price realization).

Year	Brand	Rupees kg^{-1} of processed black tea		
		Non-organic	Organic	Percent of increase
1990	Darjeeling	90	270	200
	Assam	45	200	344
	South India	35	NA	NA
2000	Darjeeling	150	300	100
	Assam	95	150	57.8
	South India	65	120	85

Source: Tea Board of India, Kolkata; NA: not available.

In Sri Lanka, the price premium for organic tea is approximately 80-150 percent (this may include the fair-trade premium) on top of the conventional price level. Despite a situation of oversupply, the price level for organic tea in Sri Lanka has remained stable for the last few years (www.sippo.ch).

It has been observed that consumers' willingness to buy organic food declines as price premiums increase. ACNielsen ORG-MARG Pvt. Ltd. and Research Institute of organic agriculture (FiBL) in Switzerland conducted a market study in 2002 on organic products in India and in export markets. On the basis of this study Garibay and Katke [4] reported that the price premiums obtained for different organic products vary greatly (10 to 50 percent) from country to country depending on distribution channels (supermarkets, restaurants etc) and current market conditions. However, an average across countries the price premium for organic tea was obtained 47 percent (trader level). The organic tea in UK enjoys a premium of +

60 percent on standard brands (Personal communication with Mr. William Gorman, Executive Director of the Tea Council, UK, 9 May, 2005). In European countries, Hamm[5] observed that no consumer is ready to pay premium prices for organic products (or taxes to support organic agriculture), if he/she is not sure whether an organic product is really organic, and if he/she cannot find organic products in the market.

A 15 to 18 percent price premium is considered adequate by most consumers. As a general rule, a 25 to 30 percent price premium still seems to be acceptable to a broad range of consumers and thus to retailers attempting to reach broader segments of consumers [6]. Piason [7] reported that retail prices for organic products in France are on average 25 to 35 percent higher than conventional foods. Kortbech [8] also suggested that producer and exporter in developing countries should not overestimate the price premium payable in the USA. The IFAD (International Fund for Agricultural Development) survey [9] of Latin American organic producers showed premia of between 22 percent, to banana producers in the Dominican Republic, and 150 percent to cacao producers in Costa Rica. Normally the U.S. organic consumer may accept price premiums of 10-20 per cent, or a maximum of 25 per cent. If the price premium is higher, the consumer will usually avoid buying organic products. Therefore, producers and exporters should base their export prices on realistic production costs and price expectations.

Yields and profitability

During the transition from chemical-based agriculture to natural system of farming, the yield levels definitely drop, sometimes precipitously. Before restoration of full biological activity (e.g. growth of soil biota, improved nitrogen fixation and establishment of natural pests predators,) pest suppression and fertility problems are common. The degree of yield loss varies and depends on inherent biological attributes of the farm, farmer expertise, the extent to which synthetic inputs were used under previous management and the state of natural resources [10]. It may take years to restore the ecosystem to the point where organic production is economically viable [11]. While analyzing yield trends after conversion to organic in six Darjeeling tea gardens, the loss of crop in the first year after conversion was found to be varied between 2 and 14 percent [12]. The depressing effect on yield was significant till fourth year and reduction was reported to be approximately 45 percent [12, 13]. Out of six gardens, only in Makaibari tea garden, the yields began to increase slowly from the fifth year as a consequence of recycling organic matter and the resurgence of effective natural enemies of the crop pests [12].

Recently, a study on the yield loss after conversion to organic cultivation was conducted in Ambootia tea estate (location: $26^{0}53'$ N, $88^{0}11'$ E / 700 m - 1400 m), one of the respected organic tea estates in Darjeeling, which was established in 1861 by the Darjeeling Tea Company of England and located in Kurseong in the Darjeeling district. In 1994, the tea estate was brought under organic cultivation according to the guidelines of the International Federation of Organic Agriculture Movements (IFOAM). The organic system of farming in Ambootia was further supported by the adoption of the biodynamic methods of farming. It has been observed that a fall in production was witnessed during the transition from chemical-based farming to organic method. It can be seen from Table 10.2 that the loss of crop in

Table 10.2. Yield trends before and after conversion to organic in Ambootia tea estate.

Year	Average yield kg ha^{-1}#	Average cost of production Rs. kg^{-1}*	% yield loss over BC	Average yield loss (%) during DC and AC
Before conversion	559•	187	-	
1994	488	191	12.7	
1995	461	196	17.3	19.1 (DC)
1996	405	200	27.5	
1997	393	205	29.7	
1998	497	209	11.1	
1999**	478	214	14.4	
2000**	434	218	22.3	
2001**	439	223	21.5	23.2 (AC)
2002***	395	227	29.3	
2003	454	232	18.7	
2004****	343	236	38.6	

Average yield of processed tea in the tea estates of Darjeeling hills is 525 Kg ha^{-1}
1 Euro = Rs 50=00; 1 US$ = Rs 46=00
Before conversion to organic production (BC); • Average yield for the period from 1989 to 1993; year 1994 to 1996 = Conversion years (DC); year 1997 to 2004 = After conversion to organic (AC) cultivation and tea sold as certified organic tea
* Including cost of inputs and processing of tea, labour wage plus other benefits extended to the tea plantation workers; ** less precipitation; *** less precipitation recorded during May and June; **** Less precipitation and severe *Helopeltis* infestation, and ambient temperature during peak flushing period was also found lower than the normal.

1 Euro = Rs 50=00; 1 US$ = Rs 46=00
Before conversion to organic production (BC); • Average yield for the period from 1989 to 1993; year 1994 to 1996 = Conversion years (DC); year 1997 to 2004 = After conversion to organic (AC) cultivation and tea sold as certified organic tea
* Including cost of inputs and processing of tea, labour wage plus other benefits extended to the tea plantation workers; ** less precipitation; *** less precipitation recorded during May and June; **** Less precipitation and severe *Helopeltis* infestation, and ambient temperature during peak flushing period was also found lower than the normal.

the first year after conversion was 12.7 per cent. In Ambootia, productivity of organic tea declined around 20 per cent during conversion. The overall production declined 23.2 percent after conversion. The depressing effect on organic tea yield was significant till 2004 and reduction was to the tune of 38 per cent [14].

This is in conformity with the findings of Mohon and Bächi [15]. Of course, no general conclusions on the yield gain and loss could be drawn from this case study. However, this might be useful to the producers who consider conversion to organic tea estate. It has recently been reported that the pre-conversion yield has only been reached by some gardens [16] but there was no net gain [13] and in general most are still experiencing declining yields. Most Darjeeling estates currently make marginal profits, so a crop loss to a higher extent is paralyzing unless compensated by the suitable price increase. It is an undisputed fact that organic agriculture must reach an economically viable threshold; else it would not make any sense. Further, yield loss after conversion to organic in some of the tea producing regions of India has also been carried out (Fig. 10.1). The loss of yield varies from 22 per cent in Assam to 64 percent in Uttaranchal. The length of time taken to recover to the original yield depends on the condition of the garden before conversion. (eg. age of bush, extent of soil degradation).

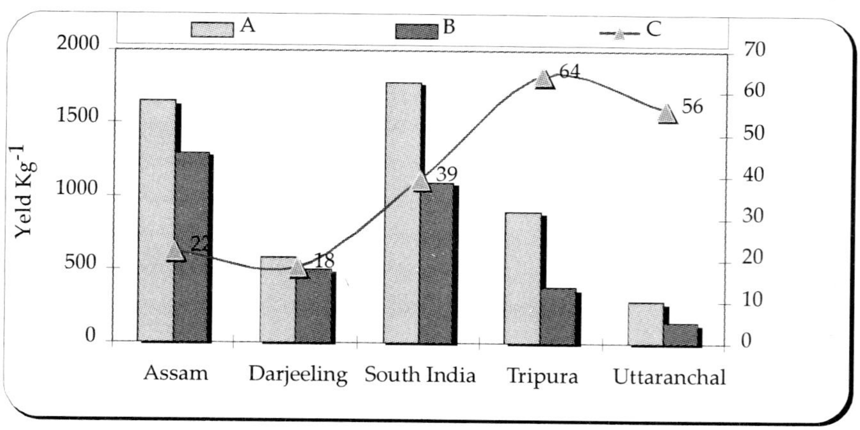

Fig.10.1. Loss of yield after conversion to organic in the different tea-producing regions of India (A= Conventional productivity, B= Organic productivity, C= percent loss in yield). Average organic yield calculated from the yield obtained from 12, 17, 15, 1 and 3 nos. tea estates for Assam, Darjeeling, South India, Tripura and Uttaranchal respectively. Source: Ref. 13..

Werf [17] reported that tea yield decreases 21 to 33 percent in the first year of transition from conventional (240 to 300 kg N/ha/year) to organic cultivation in South India. Sudden withdrawal of inorganic fertilizers might be the reason for depression in yield. But there are few exceptions where yields in organic agriculture are comparable in both systems. In the medium term and depending on new knowledge, yields improve and the systems' stability increases. In the longer term, performance of organic agriculture increases parallel with improvements in ecosystem functions and management skills [11]. Fukuoka opined [18] that once stability in ecosystem has been achieved, agro-practices adjusted/refined, the yield compares with chemical-based farming system. Further, studies have shown [19] that the common agricultural combination of lower input costs and favourable price premiums can offset reduced yields and make organic farms equally or often more profitable than conventional farms.

Much evidence has been published in recent years that contradict the conventional wisdom that organic farming leads to a decrease in yields. One of the most extensive studies in organic systems in the developing countries has been carried out by Pretty and coworkers [20] as part of a broader review of sustainable farming systems. Their results showed that the conversion to organic farming has increased yields of different crops between 30 and 500 percent. This led them to the conclusion that:

"in all cases where reliable data has been reported (from organic systems) increases in per hectare productivity for food crops and maintenance of existing yields of fibre have been shown. This is contrary to the existing myths that organic agriculture cannot increase productivity"

Recently, a stochastic analysis on the technical efficiency of organic smallholding sector in the mid country wet zone of Sri Lanka has been carried [21]. The results indicated that there was great potential to increase production by 55 percent through efficient use of the present available technology. The results identified a significant relationship between compost application and labour on organic tea production. A significantly positive relationship between technical efficiency and training of farmers was also found. But the characteristic crop diversification observed in the organic farming system has significantly reduced the efficiency.

Conversion period

It is an intermediate phase when attempts are being made to

rebuild the soil ecosystems and neutralization of the chemical residue left in the soil, to make suitable for organic management called the conversion period. The time required to achieve these goals largely depends on the pre-existing ecological condition of the soil, which in turn is determined by the past land use. Further, the conversion time is required, first of all, to begin the task of planting grasses and other bio-pest repellents around the garden. Secondly, to ensure enough manure, it is necessary to keep adequate livestock. Thirdly, a number of herbs should be grown along with tea to ensure adequate mulching and to fertilize the soil. These are aimed at creating an optimal soil condition for an enhanced biological activity in the soil so that plants get fed through the soil ecosystem and not through chemical fertilizers added to the soil. The process of rebuilding invariably demands some time. In fact, these time-consuming changes at the soil level are among the prime reasons for requiring an interim period for conversion prior to the certification as an organic unit. The conversion period may not always be of identical duration for different situations. Although the rules of the most certifying organizations require a conversion period of at least three years, the certifier has the discretion to extend the conversion period depending on the ecological conditions of the tea estate undergoing conversion. For newly-planted or replanted area raised through organic cultivation practices, the first year itself can be considered as organic produce provided chemicals have not been used in the previous cropping. In the case of cultivation on virgin land and for estates where records are available that no chemicals were applied previously, the conversion period can be relaxed by the certifying agencies.

References

1. Lampkin, N. 2002. *Organic Farming*. Old Pond Publishing, London, p. 747.
2. Das, K. 2004. Transitioning to organic. Downloaded in June, 2004 from www.indiatogether.org
3. Stamp, K. 2001. Tea – a fair cup. *Fair Trade Year Book 2001* (Revised by Claudia Grözinger), In: www.eftafairtrade.org/Document.asp
4. Garibay, S. V. and Katke, J. 2003. Market opportunities and challenges for Indian organic products. *FiBL and ACNielsen ORG-MARG*, 1-38.
5. Hamm, U. 2001. Promotion strategies and arguments for organic food in European countries. *Proc. Conf. Organic Food and Farming*. May 10-11, 2001, Copenhagen, Denmark. The Danish Ministry of Food, Agriculture and Fisheries. p. 187-194.

6. Dabbert, S., Haring, A. M. and Zanoli, R. 2004. *Organic Farming: Policies & Prospects*. Zed Books, UK. p. 169.
7. Piason, Frank J. 1999. France - Organic Food report 1999. GAIN Report FR 9070. Downloaded at http://www.fas.usda.gov/gainfiles/199910/25545926. pdf
8. Kortbech O.R. 2002. The United States Market for Organic Food and Beverages. *International Trade Center*. UNCTAD/WTO.
9. IFAD 2003. The adoption of organic agriculture amongst small farmers in Latin America and the Carribean: thematic evaluation. *IFAD Report* Rome, Italy. p.1337.
10. FAO. 1999. *Organic Agriculture*. Committee on Agriculture 15th Session, 25-29 January, 1999. COAG/99/9. Rome.
11. FAO. 2003. *World Agriculture: Towards 2015/2030*. An FAO perspective (ed. Jelle Bruinsma). Earthscan Publications Limited, UK and USA. p. 431.
12. Ghosh Hajra, N. 2002. Organic tea growing in Darjeeling hills: A case study. *Proc. PLACROSYM XV*. 10-13 December, 2002. Mysore, India, 437-447.
13. Ghosh Hjara, N. 2004. Organic tea cultivation: present state and future prospects. *Proc. Sci. Seminar on Darjeeling Tea*. Sept. 27, 2005, Darjeeling, Tea Board, India, 5-24.
14. Ghosh Hajra, N. 2005. Organic tea growing in India: a case study in Darjeeling hills. *Proc. National Workshop on Vision & Strategies for Tea Development in Hilly States*. May 28-30, 2005, Uttaranchal Tea Dev. Board, Almora, Uttaranchal, in press
15. Mohon, B. and R. Bachi. 1999. Organic tea. In: Global Advancement in Tea Science (ed. N. K. Jain). Aravali Books International, New Delhi, India. p. 813-816.
16. Jha, V. 2004. India. In: *Trading Opportunities for Organic Food Products from Developing Countries* (eds. René Vossenaar and Els Wynen). United Nations Conference on Trade and Development (UNCTAD), Geneva, p. 163.
17. Werf, E. J. V. 1990. Organic tea cultivation in Singampatti group of B.B.T.C. India. ETC Foundation, Leusden, The Netherlands.
18. Fukuoka, M. 1990. *The One Straw Revolution*. Friends Rural Centre Rasulia, Hoshangabad, Madhya Pradesh, India.
19. Rai, M. 2005. Organic farming: potential and strategies. 14th Millennium Guest Lecture at S. V. Agricultural College, 3rd June, 2005, Tirupati, Tamil Nadu, India, 1-13

20. Pretty, J., Morrison, N. J. I. L. and Hine, R. E. 2002. Reducing food poverty by increasing agricultural sustainability in developing countries. *Agriculture, Ecosystems and Environment* **95**: 217-234.

21. Jayasinghe, J.M.J.K. and Toyoda, T. 2004. Technical efficiency of organic tea smallholding sector in Sri Lanka: a stochastic frontier analysis. *Int. J. Agric. Resources, Governance Ecol.* 3(3/4): 252-265.

Chapter 11

Marketing of Organic Food and Beverages

The mechanism of organic marketing is quite different from that of regular marketing (of produce of conventional farming). Organic markets are still a niche segment in which specific buyers are targeted. Careful selection and development of target markets and distribution channels is of utmost importance in case of organic. Such marketing requires different skills than regular marketing and may call for additional costs in the initial stages.

Market and sales in US, EU, and other Countries

Retail sales of organic food and beverages in European countries have enjoyed substantial growth over the past ten years. Sales growth in some countries has been as high as 85 percent. However, it was overtaken by North America (11,000-13,000 million US$) during 2003 followed by other countries like Canada (850-1000 million US$), Japan (350-450 million US$) and Oceania (75-100 million US$). The annual growth rate is expected to reach 15-20 percent in USA over the past few years compared with 2-3 percent in the conventional food industry. In absolute terms, retail sales during 2003 are substantial and stand at around US$ 23000 to 25000 million for the most important countries (Table 11.1). However, as a percent of overall food market shares organic products continue to be a small percentage of the total[1].

U.S. retail sales of organic foods and beverages, which have grown approximately 20-24 per cent per year for the past 12 years, and reached slightly more than $13 billion (US) during 2003 [2]. Three forces aided this substantial growth: rising consumer concern about the integrity of the food supply; governmental standards that clearly define the meaning of the term "organic;" and greater availability of organic foods through mainstream channels (Mintel Report "Organic Food and Beverages - US - August 2004"-http://reports.mintel.com/sinatra/reports/display/id=112796#about).Packaged Facts, a market research firm indicates in their recent report "The U.S. Market for Organic Foods and Beverages" that the market for organic food and beverages in USA continue to grow at a rapid pace and is expected to

Table 11.1. Overview of world markets for organic food and beverages.

Markets	Retail sales 2003 (million US$)	Percent of total sales	Expected annual growth Rate (percent) 2003- 2005	Retail sales forecast 2005 (million US$)
Germany	2800-3100	1.7-2.2	5- 10	-
U.K	1550-1750	1.5-2.0	10-15	-
Italy	1250-1400	1.0-1.5	5- 15	-
France	1200-1300	1.0-1.5	5-10	-
Switzerland	725- 775	3.2-3.7	5-15	-
The Netherlands	425-475	1.0-1.5	5-10	-
Sweden	350-400	1.5-2.0	10-15	-
Denmark	325-375	2.2-2.7	0- 5	-
Austria	325-375	2.0-2.5	5-10	-
Belgium	200-250	1.0-1.5	5-10	-
Ireland	40- 50	<0.5	10-20	-
Other Europe*	750-850	-	-	-
Total Europe	10000-11000	-	-	-
U.S.A.	11000-13000	2.0-2.5	15-20	-
Canada	850-1000	1.5-2.0	10-20	-
Japan	350-450	<0.5	-	-
Oceania	75-100	<0.5	-	-
Total	23000-25000.	-	-	29000-31000

*Finland, Greece, Portugal, Spain, Norway, Poland, Hungary, Czech Republic, Estonia, Lativia, Lithuania

Source: International Trade Centre, compiled in 2002; *World Organic News*, 27 Feb, 2003, No. 64.

generate sales of US$ 32.3 billion by 2009, although other sources are less optimistic and think it will take longer to reach this level [3]. The U.S. is the world's largest market for organic products and the main factors in this positive growth are strong consumer awareness of health and environmental issues[4, 5]. The organic food is sold to consumers in USA through the following channels [6]: • natural food stores (48 per cent); • conventional grocery stores (49 per cent); • direct-to-consumer markets (3 per cent). This contrasts sharply with the early 1990s, when an estimated 7 percent of all organic products were sold in conventional supermarkets and 68 percent in natural food stores.

The U.S. standards require mandatory certification by a USDA-accredited certification agent for products labelled "100 percent organic," "organic," or "made with organic" on U.S. market shelves. Anyone claiming a product is organic without certification to U.S. standards will be subject to a fine of US$10,000 or a year in prison. Producers selling less than US$5,000 worth of organic products a year are exempt from becoming certified; however, even they must follow provisions of the national organic standards.

The U.S. also has numerous organic traders, brokers, wholesalers, processors, many of whom are members of the OTA (Organic Trade Association) through which they may be contacted. Direct sales from exporters to the central purchasing units of retail chain stores/ supermarkets are still rare, but this may change over time.

The European market for organic food is undergoing strong expansion[7]. However, its development varies from country to country. By the year 2005 the share of organic foods in the European market is estimated to reach values of between 5 and 10 per cent [8]. In the EU, for each market and for each product group, a few specialized organic traders tend to dominate imports. Personal relationships are important, and establishing a long term relationship with one of these companies is a key requirement if exporters from developing countries wish to gain entry into the EU market.

In developing countries, organic production is mostly export-oriented and the main parts of production are marketed through exporters to the main markets such as the EU, the USA and Japan. Generally, there is a simple chain between producers and processing companies or products are directly distributed to importers. The marketing channel in the EU is presented in Figure 11.1.

With the foundation of the Soil Association in 1946, organic farming has a long history in the United Kingdom. It is only since the late 1990s that demand for organic products has risen significantly. This is due in part to the advertising efforts of the retail trade. The British retail market growth for organic products has been one of the fastest in Europe. The retail sales in 2003 was reached around Euro 1750 million[9]. Estimates of the share of organic products in the overall retail market for the year 2002 ranged between 1.5 - 2.0 percent. The main organic products are fruits, vegetables and herbs. Tea and coffee are 100 percent imported. In the British organic food market, multiple retailers are responsible for 80 percent of the total organic food sales,

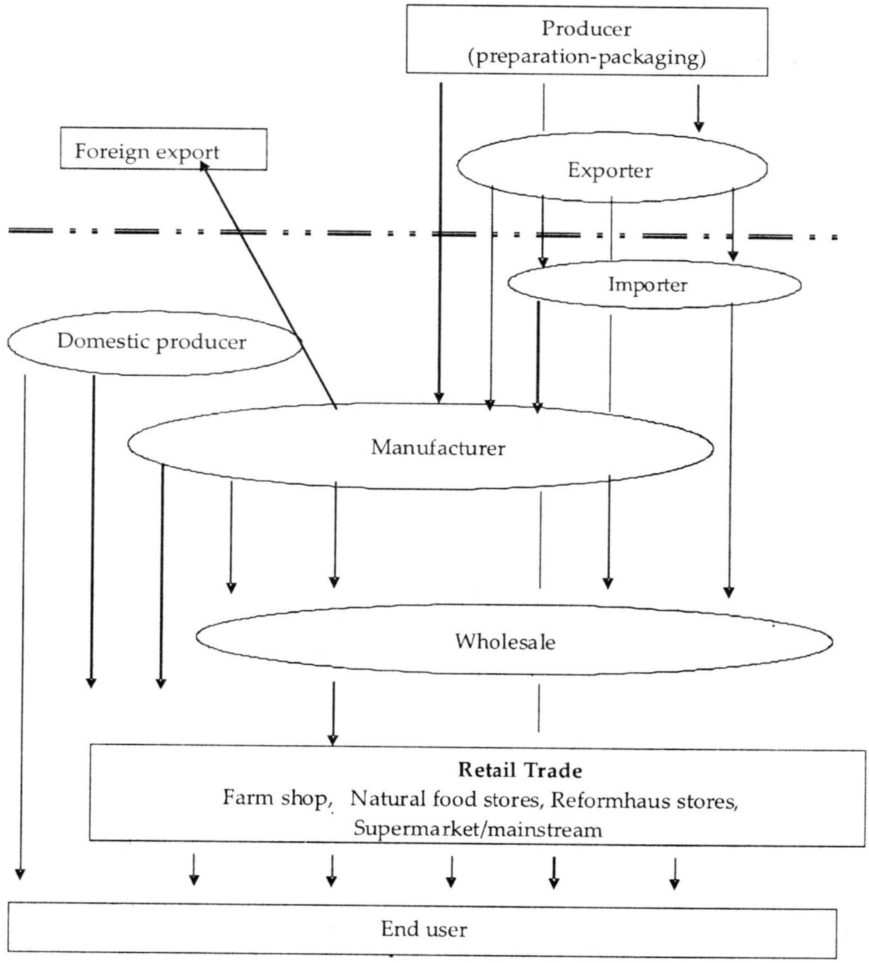

Fig. 11.1. Marketing channels for the European Union. Source: Ref. 12.

9 percent is done through local and direct sales and 11 percent through independent and health food shops [10]. Box schemes, involving delivery to the home, directly or by mail order of a box of fresh in-season organic fruits and vegetables, are a significant market outlet. Experts consider annual growth in sales by 10–15 percent to be realistic, reaching a value of Euro 2.0–2.5 billion by the year 2005, which would amount to 2 to 3 percent of total food sales. Trade estimates even put the market share at 10 to 15 percent by the year 2009 [9].

In the United Kingdom, the market access for organic products is regulated by EU Regulation 2092/91 on organic farming. The Swiss Co-op chain and Sainsbury's in the UK are the supermarket chains with the highest sales of organic products. In UK, Switzerland and the Sweden supermarkets account for well over 50 percent of the organic trade - and their share is increasing. The density of organic retail outlets, for all countries, is strongest in the cities - where there are a large number of affluent consumers. Five retailers (Sainsbury's, Tesco, Asda, Safeway and Somerfield) are found to be dominated in the UK market [11]. Traidcraft plc, one of several fair-trading organizations (may see Box 3.1) in the United Kingdom, sells Darjeeling tea which is certified organic at source. Only about 2 per cent of Traidcraft's turnover is derived from organic foods. Oxfam, too, sells tea with the fair trade mark but without an organic label, though the tea estates it buys from are converting to 100 percent organic gardening under IFOAM guidelines. Clipper Fair Trade Tea has also had a degree of market success since its launch in 1994. Another fair trade organization, Equal Exchange, supplies wholesalers with retail packs of various teas (from India and the United Republic of Tanzania) with the dual fair trade/organic tag.

Germany is by far the largest market in Europe and the second largest in the world after the United States. 38 percent of the product value of all organic products is imported. Fruits and vegetables (fresh and processed) are the largest import items, with a share of 30 percent in terms of turnover of all imported organic goods and a share of 50 percent of organic fruit and vegetable consumption. In organic cereals and cereal products, and in eggs and poultry meat, imports account for 10 to 20 percent of domestic demand. Important organic import items also include tea, nuts, coffee, cocoa, spices, culinary oils and fats, sweeteners and bakery products. A large proportion of imports come from emerging markets and markets in transition [9]. Market access for organic products is mainly regulated by EU Regulation 2092/91 on organic farming. Germany accounts for about one third of total sales followed by France, the United Kingdom, the Netherlands, Switzerland, Denmark and Italy (in Italy, supplies are mainly of domestic origin). Organic food sales in 2002 was Euro 3.00 billion in Germany. This corresponds to a share of 2.3 per cent of total turnover of the food market. By 2005, the turnover is expected to double [9]. Germany would seem to be the country that offers the most opportunities for organic agricultural products because of its size; an environmentally sensitive population, with very high purchasing power and a long tradition of natural foods and medicines. Furthermore, Germany has well-organized promotional programmes for organic agricultural products. The largest fair in organic agricultural products, BIOFACH, is organized annually in Germany, while Protrade/GTZ runs the GreenTradeNet Web site, which is an important tool for linking organic farmers in developing countries with the German market. Supermarkets are dominated by 'discounters' (e.g. Aldi and Edeka) - these retail on price rather than added value and as such the majority of organic sales are passed through the specialist retailers (Reformhaus, Naturkost and independents) in Germany. The large multiple retailers such as REWE, Metro and Tengelmann in Germany have introduced organic lines and are also showing significant growth. However, health and natural food stores still dominate today in Germany (35 per cent of total sales of organic products). Only 35 percent of all organic products are marketed through supermarket chains today while further 18 percent are sold via direct sales [7]. Nonetheless, German supermarkets are expected to increasingly enter the organic market. The specialized organic supermarkets that opened in the late 1990s are an attractive supplement to the specialist organic retail trade. Catering also has growing demand.

The Netherlands is a significant gateway for potential exporters from developing countries interested in penetrating the North European market. Dutch traders are highly specialized importers and re-exporters of bulk organic produce, especially non-seasonal fresh fruits and vegetables, tropical fruits, preserved fruits (dried, concentrates, purees etc) grains, nuts, coffee, tea, herbs, spices etc. They typically work very closely with producers from the very early production stages to the final stage of distributing to consumers according to the standards of the EU and national regulations. Market growth in The Netherlands is driven by the supermarkets. In 2001, 42.3 percent of the total sales of organic products was sold via supermarkets against 40.8 percent in whole food stores i.e. health food shops and natural food stores and 16.9 percent was sold through other distribution channels like bakeries, butcheries and direct selling. Most supermarket chains in The Netherlands such as Konmar, Vomar, Dekamarkt and De Nieuwe Weme sell organic products in their branches [9].

The French organic market began to develop in the 1970s. In the 1990s, growth in both organic production and consumption accelerated, but at a much slower rate in France than in many European countries. France has a market share of 1.0–1.5 percent of total food sales. Growth rates of around 5–10 percent are expected for the organic market in the coming years [9]. Euromonitor reviewed (at: http://www.euromonitor.com/Hot_Drinksin_France)French market for hot drinks and showed that the best performing sector was tea with volume and value CAGRs (compound annual growth rates) of 4.8 per cent and 6.5 per cent respectively and overall growth of 26.4 per cent in volume terms and 37 per cent in current value terms. It was reported that ethical and environmental issues are becoming a selling point and more products are being introduced, such as Malongo's range of organic and fair-trade tea, which was introduced in 2003. Most companies involved in the French hot drinks market are subsidiaries of multinationals. Unilever continued to lead tea in 2002. Unilever accounted for around 40 percent of tea value sales in 2002, showing little change over the 2000-2002 periods. The supermarkets/hypermarkets channel still accounted for around an 83 percent volume share of retail sales in 2003 (at: http://www. euromonitor.com/Hot_Drinksin_France). Consumers of organic products in France fall into three main categories: the politically or ideologically motivated who are concerned about the environment and animal rights; the health conscious; and the switchers, who are easily swayed by the media and influenced by price and availability. About half of consumers of organic foods are managers or self-employed professionals between the ages of 25 and 49 [13].

The organic food and beverage sector in Switzerland has increased steadily its share of the food market. According to Bio Suisse (Swiss Association of Organic Agricultural Organizations) the market has experienced a growth rate of 6 per cent from 2002 - 2003. Total turnover in 2002 was US$ 782 million (at: http://strategis.ic.gc.ca/epic/internet/inimr-ri.nsf/fr/gr118646f.html). Three-fourths of this turnover occurred in Coop and Migros, Switzerland's most important chain stores. Both had a growth of 20 per cent with their organic product programs. Swiss consumers buy organic food and beverages for a variety of reasons. According to a survey published by FiBL in 1998, the main reasons are environmental and health concerns, followed by taste and other factors. Usually, a price premium of 10 - 20 percent is accepted, though this may be higher for some products, say up to 40 percent. According to another survey, 40 percent of Swiss consumers buy organic food at least once a week (Table 11.2).

Table 11.2. Consumer share and price premiums in key demand centers in Europe, Japan and USA.

Market	Consumer share (% Buying regularly)	Price premium (% Above conventional)
Austria	20	25-30
Denmark	32	20-30
France	10	25-35
Germany	32	20-50
Italy	4	35-100
The Netherlands	5	15-20
Sweden	15	20-40
Switzerland	40	10-40
United Kingdom	25	30-50
Japan	4-36	10-20
United States	9-19	10-30

Source: Ref. 14

Gastro-Facts, a web-based gastronomic platform, reported that consumption of bio-food is no longer linked to those of upper incomes. The Swiss bio-product market is located mostly in German-speaking Switzerland. Surveys show that women and people with higher education buy bio products and approximately ¾ of the whole bio-food production is consumed in the Swiss German region of the country, rather than in the Swiss French or Italian regions (at: http://strategis.ic.gc.ca/epic/internet/inimrri.nsf/fr/gr118646f.html). BIO SUISSE certification plays an important role in the marketing of organic products in Switzerland; however, a product that complies with the Swiss Organic Farming Ordinance can also be marketed as organic.

On an average the *per capita* spending on organic produces in Europe was euro 23 for 2000. In terms of *per capita* consumption of organic products countries like Denmark (72 Euros per head), Switzerland (Euros 68), Austria (Euros 40) and Germany (Euros 31) fared much better than others. The absolute retail sales figure in EU countries depicted in Fig. 11.2 used in this comparison does not depend only on the size of the country and its population but also on average *per capita* spending on organic food.

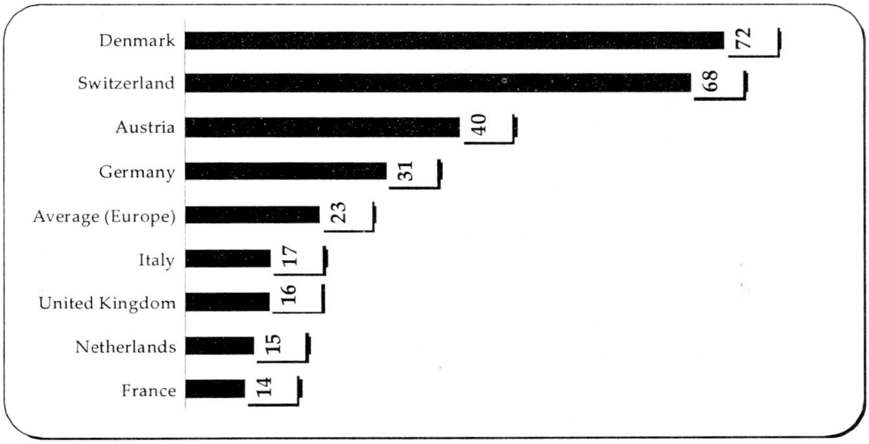

Fig.11.2. Per capita sales (in Euro) of organic produce in selected European countries in 2000. Source: Ref. 15

Danish consumers spend more on organic food than consumers in any other country. While some of the European countries like Italy not only meet their internal demand for organic products but they also cater to the demands of other neighbouring countries. There are other countries like the United Kingdom, which are highly dependent on imports to meet their domestic organic product demand.

The average consumer expenditure for organic food in European countries is presented in Fig. 11.3. Richter [16] reported that Switzerland can be considered the clear organic market leader in Europe, or even the world, with an expenditure of more than 100 Euro per capita per year for organic food.

The marketing of organic products needs to take consumers' motives into account for buying organic products. Consumers are not necessarily motivated by an adherence to organic principles; their choice in favour of organic products is more likely to reflect their general values, ethical standards, levels of personal satisfaction and well-being [17]. For example, Müller and Hamm [18] showed that in Germany aspects such as taste or animal welfare are more important motives for buying food than organic principles, although animal welfare is stated as a principle in organic standards. Surprisingly, the most important retailers' arguments for marketing organic food in Germany are quite different: environmental protection, food safety and health, taste and nature conservation [19]. In contrast, retailers in

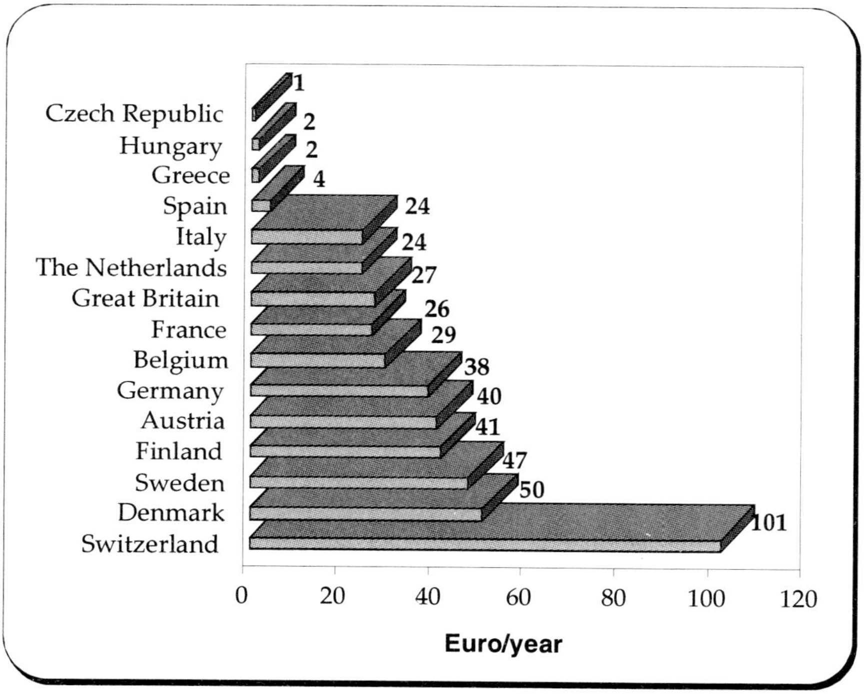

Fig. 11.3. Average consumer expenditure for organic food in European countries (Source: Ref. 16)

Italy appear to promote organic food with relevant arguments. Zanoli and Naspetti [20] showed that the most important motives for buying organic products are linked to health and well-being; the most commonly used retailers' arguments in Italy are food safety, health and taste [18]. A similar overlap between consumers' motives for buying organic food and retailers' sale arguments is observed in the UK [21].

Consumers of organic food are mainly demotivated by high prices, poor product distribution, little obvious difference in quality, a lack of information on the nature of organic products, and doubts about their genuine organic origin [20, 22]. However, the perception of organic food and farming, and consequently the motivation and demotivation of consumers to buy organic food, can change quickly in response to such factors as marketing campaigns from the organic sector or food scares [23].

Although there are wide country variations in the European consumers, a greater and more committed participation of

Box 11.1

Web Information Corner on the Organic Market in Europe

http://www.soel.de

The German Foundation Ecology and Agriculture (Stiftung Ökologie & Land- bau, SÖL) has published a study entitled "Ökologische Agrarkultur Weltweit– Organic Agriculture Worldwide" on its website. A copy of this bilingual (German/English) study may be obtained either direct from SÖL or downloaded from the internet (www.soel.de/ oekolandbau/weltweit.html).

http://www.intracen.org/

The website of the International Trade Center (ITC) in Geneva. The ITC has produced an overview of the market for organic products worldwide.

http://www.organic-europe.net

It gives reports and database on organic agriculture in Europe.

http://www.green-tradenet.de

Information network on organic raw materials worldwide.

http://www.biogene.org

Information Turntable for GM-free Production.

http://www.organicTS.com

Information about market opportunities world wide for organic products.

http://www.eisfom.org

European information System for organic markets gives information of 32 European countries for organic products.

supermarkets and chain stores has been one of the main contributors to the rapid increase in retail sales of organic agricultural products. In fact, supermarkets have shown significant commitment to organic agricultural products, by ensuring a continuous supply of organic foods, as well as a complete assortment of products. Furthermore, the growing supply and more efficient distribution of large quantities helped to lower price premia to levels more acceptable to a larger segment of consumers. Supermarket chains were also instrumental in promoting the sale of organic agricultural products through aggressive advertising. The catering and institutional use of organic foods in the EU has commenced in some European markets buthas not developed extensively to date. In North European countries, notably in Denmark, some municipal and other public bodies have started to require schools, hospitals, universities, homes for the aged to offer organic menus, including organic tea and coffee. Similarly, restaurants specializing in organic food are increasing in number. Some airline companies, e.g. Lufthansa and in particular Swissair offer organic in-flight meals (at: http://www.Unescap.org/tid/publication/chap3_2120.pdf).

The most important market segment in developed countries on which exporters from developing countries should focus is the food processing and repackaging sector. Almost all organic food and beverages imported from developing countries consist of fresh produce or raw material that needs some form of repackaging/ processing before it can be offered for sale to the consumer.

The largest Asian market for organic products is found in Japan[5]. The Japanese organic market has been characterized as a market with high demand for organic products and strong purchasing power, and with low domestic supply of organic products [24]. Growth is being driven by factors like consumers' fears for food safety and growing consumer interest in natural products.Growth in the Japanese market is also driven by the marketing efforts of the major retailers. Large supermarkets like JUSCO and Daiei have introduced organic foods in their stores in response to rising consumer interest. Convenience stores like 7-Eleven and Lawson's are also marketing organic foods. Kinokuniya, an upscale supermarket known for high-quality imported products, offers a wide range of natural and organic products. Much of the expansion in the Japanese organic food market, however, is not consumer-driven. The Ministry of Agriculture, Forestry and Fisheries (MAFF) enforced regulations for organic food production and marketing in April 2001. This allows only organic foods that meet Japanese Agricultural Standards (JAS) to be labelled and marketed as organic foods in the Japanese market (may see Box 12.3). The bulk of organic foods sold prior to the ruling lost their organic status in 2001 with the market size shrinking from an estimated US $3 billion in 2000 to about US$ 250 million in 2001 (source: *The World of Organic Products* by Heinz W. Kuhlmann; downloaded at http://www.abc-language-media-tradefairs.jp.com). The market size is expanding as more Japanese and imported products receive JAS accreditation. This is a major factor causing the Japanese market to expand and it will continue to be an important driver over the short term of the forecast period. The agricultural promotion council of Japan made a survey during April, 2002 to November, 2003 about organic products targeting 157 supermarkets and retail stores. 70 percent of them wish to increase stock of organic and low chemical farm products in their stores. However, they were found to concerns over the high prices and 50 percent of them requested supply stability of the products (Source: IFOAM Japan Letter).

A hyper-saturated coffee market has helped create the resurgent interest in tea for Japanese consumers, and over the last three years, Japan with a substantially smaller population than the U.S., has nearly equaled it in terms of the total number of products launched with tea as an ingredient. This is one of the reasons tea producers in India, China and Sri Lanka have targeted the U.S. and Japan for their renewed marketing efforts to counter supply gluts and sluggish prices being fetched in many traditional markets such as the Middle East, former Soviet Union and parts of Europe.

China, which has the largest area under organic farming, is showing a large rise in sales of organic foods because of tangible increase in production. The majority of exported products were organic tea, soyabeans, green beans, and other vegetables. In 1990, organic green tea produced in Zhejiang province was certificated by the SKAL and exported to Holland, which was initiation of organic

Box 11.2

Organic Food Market Restraints in Asia

- Many Asian countries have large sections of the population below the poverty line, and they struggle to afford the basic necessities. Organic foods are considered as luxury items mainly because of the high price premiums.
- Low consumer interest and/or awareness of organic foods—this is either because most Asian consumers do not know what organic foods are or because they are not interested to buy organic foods.
- Low production levels of organic foods in the continent, which impacts availability and consumer awareness. These are the reasons why demand for organic foods is limited to the most affluent countries like Japan, Hong Kong and Singapore where consumers are more informed of food issues and they have the purchasing power too to buy organic foods. Apart from Japan, organic foods are imported by all these countries because there is no domestic production.
- There are no regulations for the marketing of organic foods in any Asian country aside from Japan. Also, there is no differentiation between organic and non-organic foods in the marketplace. However, in countries like China and South Korea there is little differentiation between organic and green foods and sale of organic foods suffer because of higher price.
- Poor availability of organic foods in the market – apart from Japan, it is rare to find retailers that offer organic foods and this is because of very low production of certified organic foods and because of low consumer interest.

food production in China [25]. Recently, organically-produced foods such as tea, honey, vegetables, grains, and milk powder sold in the domestic market of China had a very fast rate of growth. In 2001, the sale value of foods sold on the domestic market with the Green Food designation (see below) reached around US$3.5 billion (http:// www.itmonline.org/ arts/organics.htm). By 2000, there were nine research groups that carried out research on organic food; nearly thirty companies undertook trade in organic foods; more than fifty bases of organic food production; and more than thirty factories that handled organic food and herbs in China (http://www.itmonline.org/ arts/ organics. htm).

India is also a major producer of organic agricultural products and as is the case for China is considered as a country with tremendous potential for further expansion. Basmati rice and specialty teas are among the most successfully exported products. The number of organic traders is growing rapidly.

Sri Lanka is one of the world's leading sources of organic tea. It also produces other certified organic products, including cane sugar, coconut products, coffee, fruits, fresh dried and canned fruits and concentrates, grains, nuts, herbs and species, oilseeds.

In Indonesia, vanilla has become an export product; while the Philippines have become an important producer of organic coffee. Thailand has significant production in grains. Fragrant organic rice has been successfully grown in Chiang Rai province. Most of the rice is exported, but it is also available on the domestic market selling at about double the price of regular rice. Demand for organic products in Asia, despite Japan, is largely confined to most of the affluent countries, notably South Korea, Hong Kong, Taiwan, Singapore and Malysia. Domestic price premiums range from 10 to 400 percent, according to market location, quality and products [26]. Organic produce is becoming mainstream in Egypt. The biodynamic SEKEM initiative, employing about 1000 people, delivers its products to 6000 pharmacies and to 1200 shops. Egypt, being a nation of tea drinkers, has shown its preference for organic tea by the fact that the best selling herb tea is certified organic. More concerted consumer campaigns combined with export promotion strategies are required in Asia for economic sustainability of the organic agriculture production sector.

Organic tea market

The range of marketing channels is diverse, as are market

conditions from rural India to Tokyo, including adhoc organic bazaars, small retail shops, supermarkets, multilevel direct selling schemes and internet marketing. In Germany the bulk of organic tea is sold via organic shops or health food stores, in the UK and the USA via conventional retailers. More recently, organic tea has also been marketed via special tea shops. The sale of excellent and expensive tea specialities is currently less important in Europe. There are numerous web sites that sell organic tea via the internet. Producers, manufacturers, distributors and various other companies are using this sales channel, and different kind of teas is available, often at lower prices than in retail shops. Traders confirm in interviews that even organic tea is affected by a certain consumer price consciousness, which hinders broader sales of high quality organic tea (www.sippo.ch). However, from the consumer's point of view, it is very difficult to judge the right price for organic tea quality amid the mass of competing brands and varieties. At least in Germany there is no clear visual or terminological indication of different production methods and tea qualities. Most consumers are only able to choose between different origins (e.g. Darjeeling *versus* Ceylon), varieties (e.g. herbal *versus* black tea) and partly between different tea gardens. However, there is no evidence about methods of production and processing apart from the general term 'organic production'. Despite the provision of general information material by the suppliers, most consumers are not able to find it easily at the point of sale (www.sippo.ch).

Considering the increase in organic farming activity in India, NGOs (Non-governmental organisations) were asked to estimate the overall growth in the market for organic products. The major organic products for which a growing demand is anticipated would be tea, spice and bananas. The growth forecast in 5 years was 13, 14 and 15 percent respectively (ACNielsen Org-Marg, 2002). In India, with the growing consumer awareness regarding food safety, health and environmental issues the demand for organic tea like other food items has also been growing rapidly since it was introduced in the late 1980's. The growth in the production of organic tea from 1990 to 2000 has been 20-fold. The production of certified organic tea reached 3,000 and 3,500 mt in 2002 and 2003 respectively, and 6600 mt (certified and under conversion) in 2003. Currently, about 75 percent of certified organic tea is being exported to Germany, France, the United Kingdom, Japan, and the United States [27]. Fig. 11.4 shows percent organic tea consumed by the importing countries.

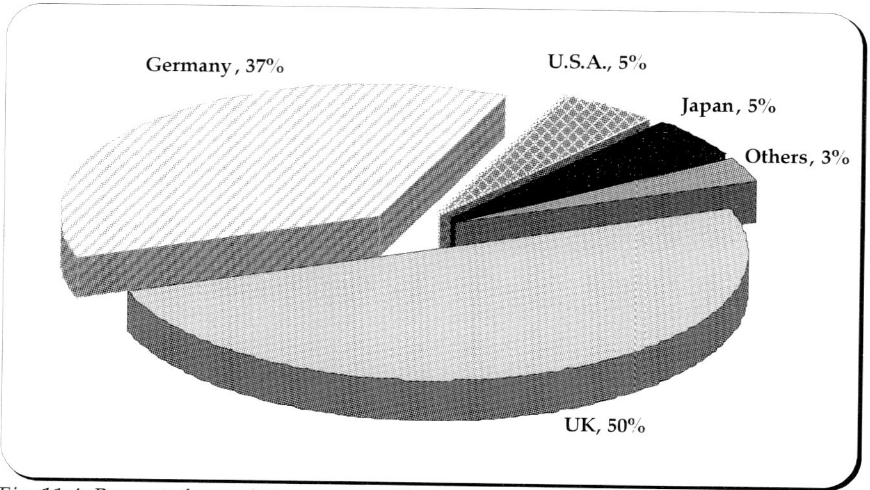

Fig. 11.4. Percent of organic tea consumed by importing countries. Internationally certified organic tea production was 12,500 tons. Source: Adopted from the data available at www.sippo.ch

India leads the world in organic black tea output and Sri Lanka is quite strong as well. The domestic market for organic tea does not yet exist in India and Sri Lanka. The market in India in fact has not grown large so far due to lack of marketing initiatives from producers, traders and NGO's, and the low awareness and higher price of the product. However, there seems to be increasing health awareness spreading among the literate population, and pesticides were broadly discussed in the media as a likely source of various health problems. It can be assumed therefore that awareness of the health benefits of consuming organic tea will increase considerably particularly among the urban upper-middle class and upper class in the coming years.

Most of the profitable enterprises have invested a substantial portion of their profits in marketing tea in their main markets for over 10 years. As the owner of a very successful tea garden put it: "A legend has to be created" in the minds of buyers. Alongside, there is brand creation. While Ambootia and Makaibari are established estate brands, more recently, a few companies have ventured into consumer packs for domestic and international markets, clearly positioned as "organic". Ambootia's largest market is Europe specifically in Germany. Ambootia Tea sells in two market niches: the organic, natural foods market and the high-end market. In England, Ambootia tea is sold exclusively at Harrods and served at the Georgian, the

Box 11.3

Trends in Organic Market

The organic food supply chain is a typical consumer driven sector, with a market value of US$40 billion (2004) and a growing involvement of the private sector. Although the devolution of the state is increasing in all fields, government regulations in organic agriculture are increasing. 60 countries are at some stage of regulating the organic sector, of which 37 are implemented. The organic market, initially led by producers and traditional organic communities is gradually being dominated by large and vertically coordinated supermarkets, in terms of market share and ownership of top brands. Supermarket sales of organic produce amount to 40 per cent in Germany, 49 per cent in the United States of America, 80 per cent in Argentina and the United Kingdom, and 85 per cent in Denmark. Natural food supermarkets are rapidly growing. Whole Foods is the largest chain in terms of sales, with 140 stores and US$ 2.7 billion in the United States. Most large food companies have acquired organic brands and small firms, set up partnership with organic companies, or have their own organic lines. Cadbury Schweppes (organic juices), Coca Cola (acquired Odwalla Organics), Danone, Dean Foods, Dole, General Mills, Heinz, Kellogg, Kraft, Mars, Parmalat, Sara Lee, TysonFoods (organic meat). Mergers and acquisitions of organic brands and companies impact production, processing, certification and distribution pathways. In California, 2 per cent of organic growers represent 50 per cent of organic sales; Coleman owns half of the United States organic pastures; Hain-Celestial is the world's largest processor of organic foods (annual revenues of US$ 400 million). Tree of Life is the world's largest organic food distributor (US$ 3.5 billion of sales); the Hain-Heinz partnership has 71 per cent of the global market share of organic baby foods. This industry concentration sets prices, limits farmers' return, leverages supply and controls market access. In 2001, Horizon Organic reported a 200 per cent increase in profit net sales (US$ 160 million) but actually cut contractually-promised premium to farmers after it bought out the Organic Cow of Vermont, decreasing farmers' income by US$ 15,000 per family. Currently in the United States, the food industry pressure on the Government is threatening national organic standards (with proposed modifications on synthetic substances, animal feed and commercial availability of organic ingredients without public review). A reality today is that "demand" is driven by big retailers with brands that dictate standards and a market economy which is anything but transparent.

Source: Nadia El-Hage Scialabba, FAO of the United Nations (www.fao.org/organicag).

Box 11.4

Export Obstacles for Selling Indian Organic Products

ACNielsen ORG-MARG Pvt. Ltd. and Research Institute of organic agriculture (FiBL) in Switzerland conducted a market study in 2002 in order to find out why traders do not import Indian organic products into the selected markets. The traders' perception of obstacles for buying organic products from India was organized according to the different aspects which are listed below.

Aspect	Obstacle
Price	Expectations are too high compared to the quality
Quality	Low consistency of quality; contaminations
Availability	Reliability of exporters, lack of information
Logistics	Slow shipment
Certification	Lack of national certification
Export authorities	Time-consuming and complicated paper work
Information	Lack of information on availability and certification
Administration	Inconvenient modus of payment (letter of credit), the bank system is too slow

Replies of questions from 33 importers in Germany, the UK, Switzerland, Japan and the USA

world-famous restaurant within Harrods. In the United States, two tea retailers now market Ambootia products in the States: Stash Tea and The Republic of Tea. Stash carries four kinds of Ambootia's black teas as well as the green tea. And three out of eight of The Republic of Tea's "Estate Tea Collection" is Ambootia products. Personal contracts are considered very important in establishing the organic tea market. Organic tea producers also have to point out all premium benefits for their products in powerful promotion campaign which facilitate to get premium prices. This also ensures that the buyers getting a quality product along with his requirements at competitive prices [28]. It has been reported that out of 30 organic gardens in Darjeeling only few are able to make profits (www.unctad.org./trade_env/docs). Those that are able to make profits have directly tied up export contracts with their buyers and export mainly to Germany, Japan and USA.

Seeing the success of organic processes in large estates, many small growers of tea in India have geared themselves for organized organic cultivation. Kerala-based groups such as the Peermade Development Society and the HICOS cooperative in Wyanad have streamlined their processes, improved their quality. In India, it is now convinced that the market for organic tea exists, what is needed is to organize growers to act together to produce and market their own products for the export markets. Training programs are needed for small growers and NGOs not only in organic agriculture methods (production, harvest and processing, basic standards etc.), but also aimed to help them understand how to sell, promote and diversify their markets and meet certification requirements, which requires good internal controls and administrative practices on plantations. In addition, it will help if either the government or tea companies take up an organic tea promotion drive in national and export markets.

According to Sri Lanka Export Development Board nearly 300 companies are involved in tea export. Half of them involve in export of black tea in bulk. Seven companies are involved in organic tea/ bio-tea exports. Organic tea from Sri Lanka is mainly exported to Germany, the UK, Australia, France, Italy, Japan and the USA.

In Vietnam, the market for organic tea is still relatively small, and so far, is concentrated in countries outside Vietnam. CIDSE, CECI, and SNV have projects to help the growers in marketing organic tea. These projects may lead to the development of a domestic market for organic teas, offering some market stability for organic producers.

Organic tea markets in China

Among the organic produce, organic tea was the first exported item from China. China has only recently gotten into gear with production of organic tea after strong advisories from international tea importers, especially Germany. Organic tea is sold in both domestic and overseas market. Organic green tea is mainly exported to Europe, the US and Japan. In domestic market, the major product is high quality green tea. The annual output is about 3000 tons and the value is about 180 million yuan [29]. In China, supermarkets in Beijing, Nanjing, Shanghai, Guangzhou, Qingdao and Wuhan are successfully selling organic tea alongside conventional products [30]. Zhangshan village in China is well known in Europe and the village accounts for 60 percent of the organic tea products on the European Union market. Wahaha (Group) Company, Hangzhou, China is the third biggest registered red-tea drinks manufacture, is now a certified organic red tea producer. Some of the large international tea companies also have joined hands with the Chinese companies to promote Chinese organic tea.

Organic tea markets in Japan

The organic tea market in Japan was 1,800 tonnes in 2001 with about 89 percent of volumes represented by domestically produced green tea. Around 200 tonnes of organic tea was imported from India and Sri Lanka. The Official Organic Certification system became mandatory in April 2001. Organic products have to be labeled with the Japanese Organic State logo (JAS) and therefore the imported organic products must meet Japanese national standards. It is likely that the 'Organic JAS Mark' is well known to Japanese consumers [5]. It is worth noting that there are large volumes of non-certified organic tea imported into Japan and this is not sold with the JAS logo.

Japanese growers supply organic green tea to supplier companies like Mitsui Ringyo. The supplier processes and packages organic green tea and then supplies either directly to retailers or passes onto distributors. The distributors are very important for the specialist retailers and the catering sector. Most sales of organic green tea are from supermarkets like JUSCO and Daiei. Organic black tea is imported by trading companies and process the tea leaves into tea bags, package, and then sell to retailers. The Japanese supplier can import directly from the country of origin, however, most companies use an importer. Table 11.3 shows market shares of the leading companies in the Japanese organic tea market.

Table 11.3. Japanese market for organic tea: Market shares of major suppliers, 2001.

	Tonnes	**Share (%)**
Mitsui Ringyo	1,500	83.3
Others	300	16.7
Total	1,800	100.0

Note: All figures are rounded; Source: Ref. 31.

Mitsui Ringyo is the market leader in the Japanese organic tea market with 83.3 percent market share in 2001. The company markets organic tea under the Nitto Koucha brand and mostly supply to the supermarkets. Its major customers are Daiei, SOGO and Ito Yakado and the Mitsui Ringyo does not sell organic tea to specialist retailers. The company is the largest supplier of JAS certified organic green tea in Japan and it imports small volumes of black organic tea from India and Sri Lanka. There are numerous other companies supplying organic tea in Japan but only a few deals with JAS certified organic tea. Two important suppliers are Takano Coffee and Bitsuwaen. Takano Coffee imports pre-packaged organic tea from India and Sri Lanka and Bitsuwaen imports organic tea from Kenya. The other suppliers were estimated to have 16.7 percent market share in 2001. Market revenues are forecast to expand from $84.6 million in 2001 to $251.5 million in 2008 with the higher growth observed in 2003 and 2004 (Table 11.4). The organic tea market was estimated to have 1.8 percent market share of the Japanese tea market in 2001 with the share projected to rise to 4 percent at the end of the forecast period.

Table 11.4. Japanese market for organic tea: Volume and revenue forecasts from 2001 to 2008.

Year	Volumes (tonnes)	Revenues ($ million)	Revenue growth (%)
2001	1,800	84.6	-
2002	2,091	97.3	15.0
2003	2,861	126.5	30.0
2004	3,468	151.8	20.0
2005	4,134	179.1	18.0
2006	4,713	204.2	14.0
2007	5,278	228.7	12.0
2008	5,806	251.5	10.0
CAGR:			16.8%

Note: All figures are rounded; Average retail price of organic tea calculated as $ 4.70 per 100 g for revenue forecasts. Source: Ref. 31.

In regard to price of organic tea in Japan, Mother's organic food shop was sold Japanese organic green tea (100g) at 510 yen (US $3.99) in spring 2002. Organic black tea from Sri Lanka and India was being marketed between 550 - 750 yen (US $4.30 - 5.87) by the organic food shop chain. In comparison, Kodawariya (SOGO) was marketing organic tea (100g) at 700 yen (US $5.48) [31].

Organic tea markets in US and EU

America is a big tea-consuming nation and it's the fastest growing tea market in the world with an annual consumption of 80-90 million kg. Americans mostly favour cold tea and Argentina is one of the leading tea exporters to the country. It is also a good market for tea bags and instant teas. Sources said that Chinese tea was prominent in the American markets but Sri Lanka's presence was insignificant.

India's presence in the market is almost negligible. In 2000, it registered its highest export to the US in recent times with a volume of only 7.47 million kg. In 2001, this dropped to around 6.5 million kg. In 2002, it went up to 7.34 million kg. However, Indian tea companies like Goodricke, Eveready and the Maud tea Co. are exporting tea to the US. The U.S. tea market was worth $5.4 billion in 2003, according to the Tea Association of the United States of America, up from $1.84 billion in 1990. According to 'The U.S. Market for Tea and Ready-to-Drink Tea', a report recently published by Market Research.com. (http://www. marketresearch.com), it is projected that the US tea market will reach US$10 billion by 2010. A major share of the U.S. retail tea revenue is generated by sales of ready-to-drink bottled and canned ice teas. With the Americans becoming more health conscious and with the health properties of green and white teas having become more publicised, tea has acquired more demand among the people, states the report. It has been opined in the said report that specialty tea sales continue to grow even as traditional black tea sales decline. Speciality tea category includes value-added, higher quality teas. The speciality tea segment also includes organic teas and Fair Trade teas. It's a hot item among U.S. consumers, who are gravitating toward more "exotic" tea blends and fueling a tea market worth US$ 6.8 billion, according to The U.S. Market for Tea and Ready-to-Drink Tea, a new report from market research published by Packaged-Facts (http://www. packagedfacts.com), a division of MarketResearch.com. Leading U.S. specialty tea brands include Celestial Seasonings®, Honest Tea®, Tazo®, The Republic of Tea®, Traditional Medicinals®, Stash Tea® and a couple hundred smaller brands and companies [32]. Today 'speciality tea' segment consists of about 5% of the market in the US. Despite the US, speciality organic teas have a good demand in countries such as Australia, West Asia and South East Asia. It is also niche market. The U.S. marketplace for organic teas is not mature and is highly fragmented. The leading organic tea brands here are generally smaller companies with annual sales under US$ 5 million. Consumer understanding of organic tea is weak. However, consumer awareness is building slowly. The certification process is getting streamlined; conventional brands are jumping into organics where they hesitated before. The market is being pioneered by smaller ventures. Sales of organic teas in teabags for many brands have enjoyed a brisk sales clip of more than 10 percent – some brands 20-plus per cent for the last one and half years. Organic loose leaf tea sales are experiencing more moderate growth of 2 to 8 per cent. This is not totally unexpected as teabags always outsell loose teas due to their obvious convenience. Black organic offerings (flavored, unflavored, breakfast blends, etc.) remain the most popular, followed by green and oolong at a distant third. Marketed under the Choice Organic Teas brand, Granum Inc., Seattle, WA, is reportedly now the single largest purveyor of certified organic beverage teas in America, offering more than 70 varieties of black, green, oolong, white teas and herbal infusions in tea bags and loose leaf styles. Choice Organic Teas was also the first tea crafter in the U.S. to Fair Trade certify all of its qualifying teas. Much of the U.S. growth in organic tea has been through incremental additions or line extensions of organic offerings by non-organic (conventional) tea brands and not full line organic tea companies.

The stronger European markets for organic teas have enjoyed years of consumer education, mandatory certification and offerings from major brands (http://www. forestrade.com/press_gourmet_retailer.htm). In Table 11.5 sale figures are given of import volumes in major markets for labeled organic tea. The current tendency towards organic convenience food in industrial countries

Table 11.5. Sale and import figures (tons) of organic tea for major markets in 2001*.

Imports	North America	Europe [3]	Japan	Total [4]
Organic tea	2000	2000	1800 [5]	5800
Fair-trade	26	1039	12	1077
Total[1]	2000	2500	No data	> 4500
Market share (%) [2]	2.2	1.2	-	-

* Source: Ref. 34.

Notes: (1) Totals do not always add up because part of fair-trade is also organic (percent organic fair-trade estimated and subtracted from total), except for orange juice, for which currently no organic fair-trade juice is available. Thus, for 2001, it is estimated that 15 percent of fair-trade bananas were organic, and 25 percent in 2002 (newly introduced fair-trade bananas are often organic).

(2) Market share is calculated as total certified as percentage of total market.

(3) EU (15) + Switzerland + Norway, excluding sales in Italy by CTM Altromercato without the FLO/Transfair label.

(4) Total including other destinations (mainly New Zealand and Eastern Europe).

(5) Including 200 tonnes export. Total 10,000 tonnes organic tea (non-JAS certified) was sold in 2001 (Ref. 31).

Sources: Organic: for Europe - SIPPO/FiBL/Naturland (2002); for United States of America and Japan: World Bank (2002) and ITC, London (2002). Fair-trade: FLO. Shade: Ref. 35.

Box 11.5

Image of Indian Tea in the Minds of North American Consumers

Tea marketers of North America are not promoting Indian tea probably because of:

- Most of the Indian teas that come here are of very mediocre quality, not at all worth the money that the seller is asking.
- Not enough Indian tea is being offered to buyers here. It's not that they are not getting samples - the tea companies here, even the small mom and pop tea shops, are getting flooded with unsolicited samples from Indian tea exporters. But Indian exporters can't expect these buyers to educate themselves about Indian tea. In fact, no one is doing that. They are sending samples and hoping that someone will order a dozen containers. If tea buyers here are offered a long-term relationship, they will probably offer Indian tea to their customers.
- Chinese and Sri Lankan exporters regularly give 150 days credit or more, but it will be hard to find Indian exporters in this market who are offering over 90 days credit. Therefore tea-marketing companies have more interest in marketing Chinese and Sri Lankan teas compared to Indian teas.
- Lack of building of good relationship between Indian tea exporters and US tea buyers, and therefore there is not much market development.
- The only exception being that there is a handful of tea marketers in the US who are doing an excellent job in marketing Darjeeling teas. A lot of Darjeeling teas easily can be seen in the market, but again, this is for only a few thousand kilograms a year.

Source: Ref. 36.

will increase, in particular for tropical beverages, baby food and frozen vegetables [33]. In Germany, 81 percent of all tea drunk is black tea, whilst green, unfermented tea holds a market share of 19 percent. Incidentally, the convenience factor has its proponents; about 40 percent of all green and black tea consumed in Germany is in tea bags. Organic tea holds a market share of 2.1 percent in 2003 (http://www.teeverband.de). India is traditionally Germany's biggest supplier of tea. Germany is also importing highest quantity of organic tea from India. In the UK, the growth of speciality tea i.e., Earl Grey, Darjeeling tea, organic tea etc is about 7 percent per annum and 5 to 10 percent of volume. Earl Grey accounts for 50 percent is worth about US$ 100 million. The organic tea in the UK can enjoy a premium of + 60 per cent on standard brands. It must be remembered, however, that the UK is a fiercely competitive market and main brands are retailing at £ 1.40/£150 for 250 grams (80 teabags) (personal communication with Mr. William Gorman, Executive Director; The Tea Council of UK, 9 May, 2005).

The Netherlands consumers drink an annual average of 83 litres of tea per person, which makes them one of the largest consumers of tea in Europe. Black tea dominates and four types of tea account for 75 percent of organic demand: Earl Grey, India/Ceylon tea, low-tannin tea and Darjeeling tea. As in the mainstream trade, demand for single-origin teas and for flavoured teas (e.g. cinnamon, forest fruit, lemon, tropical fruit, lavender, lemon grass, aniseed, etc.) is growing. A high proportion of tea is purchased as tea bags (packed in boxes containing 10 or 20 tea bags of 1.75 grams). Loose tea is packed in 100 gram bags. Tea is mainly imported loose in chests, but there is some growth in imports in consumer packing (from India and Sri Lanka). The efficiency and output of existing suppliers of organic tea are improving rapidly, according to The Netherlands trade sources [37]. These sources say that the best opportunities for first-time suppliers are likely to be in new and unusual varieties.

In Sweden, organic tea imported from Sri Lanka and India has been on the market for a few years, the market share remains small, under 1 percent [37]. Flavoured and herbal varieties of organic tea have recently been introduced. The retail market is dominated by three integrated retail and wholesale groups - ICA, Gröna Konsum (KF) and Hemköp (D&D). These have 70 percent of the food retail sector. All three groups promoteorganic and other environmentally-oriented products. It is generally believed that environmental awareness among consumers will continue to be high in Sweden.

A market study carried out in 2002 shows that in Europe the structure of the sales channels is different from country to country [7]. Supermarkets have an important role in the Scandinavian countries, Switzerland, Austria and the UK. However, in Germany, Holland, Italy and France organic products tend to be distributed to consumers via health food stores. It was further observed[7] that sale of organic products in supermarkets is growing fast and have an important role in the further development of the selected export organic market.

There is no doubt that the world markets for organic food and beverages will continue to offer developing countries profitable export opportunities. According to the ITC Organic Food Report [37], there are sound market opportunities for developing countries in most major markets especially for the products that are not produced in Europe, North America or Japan. All of the major markets under review offer good prospects for suppliers of organic products that are not produced domestically: examples are tea, coffee, cocoa, spices, tropical fruits and vegetables and citrus fruits. The certified organic tea market reached sales of US$ 7.6 million in 2000, a growth of 29 percent over the previous year. The market is poised to reach US$ 125 million by 2005 [38]. Such opportunities exist not only for off-season produce (such as fruits and vegetables), but also for many other products like in-season fruits (e.g. apples and pears) and vegetables, cane sugar, grains, cereals, pulses and seeds, for the simple reason that the rapidly growing demand in most markets cannot be met by local supplies, at least in the short and medium term. On the other hand, some multinational companies such as McDonalds, Danone, Lufthansa, Swiss Air, Nestle and Novartis have already entered the organic market [39].

Market demand for organic agriculture products has created incentives to change agricultural production policies towards a market-oriented and a more sustainable direction. Organic agriculture offers a specialized market and an opportunity to diversify into new commodities that have a high demand and price premiums. It was argued that profitability of organic methods usually depends on price premiums [40]. An analysis of the Organic Market News and Information Service shows that between May 1989 and February 1990 the whole sale price for organic lettuce ranged from 28 to 256 percent. The price premium for the red cherry tomatoes ranged from 15 to 281 percent during the same period [41].

Because of the export-oriented market structure of the organic produce, development of market capacity mainly depends on foreign import companies and their domestic branches or partners. To improve the access capacity, organic producers have to act together under their own organization such as cooperatives. In developed countries, organic standards have been developed over a period of 25 years. The main forces in this development have been the organic producers themselves. Lately, governments have also become engaged in organic standards. Having high standards, on the contrary may create trade barriers for the developing countries to some extent.

Another problem is inadequate market information (for example, which products to grow, which markets and distribution channels to choose, competition and market access) and insufficient financing. Reliable market information is almost always difficult to obtain. In particular, no projections and systematically identified markets for the countries' exports have been available in the developing world [40].

Mention can be made of fair trade practices when discussing marketing of organic products. The fair trade movement started 25 years ago. The terms of trade developed very much in favor of a developed country at that time. The fair trade started to counter this development in establishing special criteria for sustainable trade with mainly smallholders, starting with products like tea, coffee and cocoa. The criteria are mainly created to protect small farmers, farmers' cooperatives and farm workers community, providing a price premium, advance payment and a long-term trade relationship [41]. Among food items currently eligible for "Fair Trade Labels" are tea, bananas, cocoa, and chocolate. The fair trade certification is different from organic certification, although 65-85 per cent of the Fair Trade imports also carry organic certification [14].

Initially the growers, traders, certifiers and retail through health food shops were very small and most of the transnational and multinational companies stayed indifferent to this market development. However, the organic food sector in the 90's had a tremendous growth due to food safety issues such as foot & mouth diseases, mad cow disease, etc. The threshold of organic going main stream was reached and subsidized transnational and multinational corporates started taking of ownership and participating in the organic food sector. Since 2000 the rate of growth in half a decade has crossed the cumulative growth rate of 25 per cent per annum, therefore, the market dynamics are changing and with the further development of bird flu and other food safety issues as well as emerging MRL issues relating to food in developed countries the organic food sector is gaining strength day-by-day. Since organically grown food is also GMO free and satisfies all basic food standards globally, it is a sector that is expected to grow exponentially in the coming decades. It is assumed that the supply of health food will have exponential growth in the demand as the economical development and GDP growth band is 6-8 per cent. It is found that up to the age of 35 is the largest age profile consumers of organic food globally (India will have 750 million people below 35 years in the year 2020).

References

1. Pilkauskas, P. 2003. Organic market in Europe. *Proc. Seminar Production and Export of Organic Fruit and Vegetable*. 3-5 Nov. 2003, Bangkok, Thailand. FAO, IFOAM and EarthNet Foundation. 13.

2. Haumann, B. 2004. *The World of Organic Agriculture - Statistics and EmergingTrends – 2004.* (Willer and Yussefi Eds.) Bonn, Germany, IFOAM, p. 167.

3. Kortbech O.R. 2003. The North American Market. *Proc. Seminar Production and Export of Organic Fruit and Vegetable.* 3-5 Nov. 2003, Bangkok, Thailand. FAO, IFOAM and EarthNet Foundation. 14.

4. Kortbech O.R. 2002. The United States Market for Organic Food and Beverages. International Trade Center. UNCTAD/WTO.

5. Yussefi M. and Willer H. 2002. Organic Agriculture World Wide 2002, Statistics and Future Prospects. Stiftung Ökologie and Landbau. Bad Dürkheim, Germany.

6. Greene C. 2002. Organic Farming and Marketing in the U.S. *Economic Research Service.* U.S.D.A. Washington, DC. USA.

7. Hamm U., Gronefeld F. and Halpin D. 2002. *Organic Marketing Initiatives and Rural Development: Analysis of the European Market for Organic Food.* School of Management and Business. Aberystwyth, United Kingdom. P. 157

8. Kilcher L., Landau B., Richter T. and Schmid O. 2001. The Organic Market in Switzerland and the European Union: Overview and Market Access Information for Producer and International Trading Companies. Swiss Import Promotion Program and Research Institute of Organic Agriculture. Zürich/Frick, Switzerland.

9. Kilcher, L., Khanna, R., Huber, B., Richter, T. and Schmid, O. 2004. *The Organic Market in Switzerland and the European Union.* SIPPO/FiBL, Zürich/Frick, Switzerland, p.132.

10. Richter, T. and Hempfling, G. 2003. *Supermarket Study 2003 – Organic Products in European Supermarkets.* Research Institute of Organic Agriculture (FiBL), Frick. p. 180.

11. Rundgren, G. 2000. Organic markets. Entering world organic trade, Bangkok, 28 Nov. 2000. Downloaded from http://www.grolink se/Resources/studies/marketBKK.

12. Buley, M., Grosch, P. and Vaupel, S. 1997. *Export of Organic Products: Marketing Manual.* Protrade, Dept. of Organic products and Fine foods (ed. Katherine Clark). German Technical Cooperation (GTZ). Second edition.

13. Piason, Frank J. 1999. France – Organic Food report 1999. GAIN Report FR 9070. Downloaded from http://www.fas.usda.gov/gainfiles/199910/ 25545926.pdf

14. Lohr, L. 2000. *Factors Affecting International Demand and Trades in Organic Food Products.* Dept. of Agricultural and Applied Economics, University

of Georgia Athens, FS 00-20. *Consumer Cooperatives*, pp. 28-32.15.

15. Garibay, S. V. and Katke, J. 2003. Market opportunities and challenges for Indian organic products. *FiBL and ACNielsen ORG-MARG*, 1-38.
16. Richter, T. 2005. The European organic market between strong growth and consolidation-current state and prospects. *Paper presented to Biofach*, Nurenberg, 24 Feb., 2005.
17. Dabbert, S., Häring, A. M. and Zanoli, R. 2004. *Organic Farming- Policies & Prospects*. Zed Books Ltd., London, UK. p. 169.
18. Müller, M. and Hamm, U. 2001. *Verbrauchereinstellungen beim Einkauf von Lebensmitteln in Mecklenburg-Vorpommern*, Neubrandenburg: Schriftenreihe der Fachhochschule Neubrandenburg, Series A, Vol. 14.
19. Michelesen, J., Hamm, U., Wynen, E. and Roth, E. 1999. The European market for organic products: Growth and Developmemt. In: *Organic Farming in Europe: Economics and Policy*, Vol. 7, Stuttgart-Hohenheim: University of Hohenheim.
20. Zanoli, R. and Naspetti, S. 2001. Values and ethics in organic food consumption. In: *Preprints of the Third Congress of the European Society for Agricultural and Food Ethics: Food Safety, Food Quality and Food Ethics* (ed. Pasquali M.), Florence, Oct. 3-5, 2001, Milan, Italy: A&Q.
21. Makatouni, A. 2001. What motivates consumers to buy organic food in the UK? Results from a qualitative study. *Proc. 72nd EAAE Seminar*, June 2001, Chania, Crete,
22. Schaer, B. 2001. Regionales Gemeinschaftsmarketing für Öko-Lebensmittel - dargestellt am Beispiel der Konzeption des Zeichens "Öko-Qualität aus Bayern". In: *Öko-Management - Studien zur ökologischen Betriebsführung*, Vol. 4, Hamberg, Verlag Kovaè.
23. Röder, E. 2002. Biolandbau in Deutschland - Aktuelle Entwicklungen in der Erzeugung und Vermarktung. Unpublished manuscript, 10 January 2002.
24. Hiraga M. 2002. Japanese Organic Market: Market opportunities and characteristics. *Bio Market Inc.* Japan.
25. Lu, Z. 2002. Production and Market of Organic Foods in China. *Paper presented in 1st Gangjin International Symposium on Organic Agriculture*, 15 Nov. 2002, Gangjin, Korea.
26. Yussefi M., Willer H. and Geier, B. 2004. Organic agriculture worldwide. *Ecology and Farming*, January-April, 2004, IFOAM, 12-14.
27. Anonymous, 2004. *Tea Commodity Note*. Secretariate of the Intergovernmental Group on Tea, FAO, Rome, Nov., 2004.

28. Ghosh Hajra, N. 2005. Organic tea growing in India: A case study in Darjeeling hills. *Proc. National Workshop on Vision & Strategies for Tea Development in Hilly States.* May 28-30, 2005, Uttaranchal Tea Dev. Board, Almora, Uttaranchal, in press.
29. Liu xin, Shangwen, F., Zhangyu, Amin, S. and Zhichang, W. 2004. The development of organic tea in Chinese mainland. *Proc. Intl. Conf. on O-CHA (tea) Culture and Sci.*, Nov. 4-6, 2004 Shizuoka, Japan, pp. 732-735.
30. Sanders, R. 2004. The anatomy of an organic village. *China Review* **28**: 1-4.
31. Organic Monitor 2002. The Japanese market for organic beverages. Organic Monitor, London, UK. p. 76. Info at http://www.organicmonitor.com.
32. Keating, B. 2005. *U.S. Tea Industry: Heating Up & Getting Hotter.* Sage Group International LLC, Seattle, WA.
33. Anonymous, 2003. *World Agriculture: Towards 2015/2030 – An FAO Perspective* (ed. Jelle Bruinsma). Earthscan Publications Limited, London. p 431.
34. Dankers, C. and Liu, P. 2003. *Environment and Social Standards, Certification and Labeling for Cash Crops.* FAO, Rome, Italy. *P.* 103.
35. Giovanucci, D. 2001. Sustainable coffee survey of the North American speciality coffee industry. SCAA/CEC. United States of America.
36. Bhattacharjee, S. 2005. There is an emerging market for good quality Indian teas in North America. Downloaded at www.indiateaportal.com in July, 2005.
37. ITC. 1999. Organic food and beverages: world supply and major European markets. International Trade Centre, UNCTAD/WTO, Geneva. p. 271.
38. Datta, A. 2003. India to go or not to go organic? *Tea & Coffee Asia* **5**(2): 20-22.
39. Geier, B., 2000. A short overview and facts on worldwide organic agriculture. *Reports on Organic Agriculture Worldwide*, IFOAM
40. Rehber, E. and Turhan, S. 2002. Prospects and challenges for developing countries in trade and production of organic food and fibre. *Br. Food J.* **104**: 371-390.
41. Hall, C., Edward, R. and Johnson, J. 2001, *A Guide to Marketing Organic Produce* (http://aggie.horticulture.tamu.edu).
42. Cierpka, T., 1999. Organic farming and fair trade: "what belongs together will grow together", *Ecology and Farming.* pp. 11-12.

Chapter 12

Standards, Certification and Accreditation

Organic farming could be considered as a production system in which farmers merely comply with certain standards. Many industrial countries, but few developing countries have national organic standards, regulations and inspection, and certification systems that govern the production and sale of foods labeled as 'organic'. Organic agriculture, broadly defined, is not limited to certified organic farms and products only. It also includes non-certified ones, as long as they fully meet the requirements of organic agriculture. This is the case of many non-certified organic agricultural systems in both developing and industrialized countries where produce is consumed locally or sold directly on the farm or without labels. However, if products are to be marketed separately, a system ensuring that products have been produced according to the organic standards must be established and a clear distinction between organic and conventional products needs to be pursued throughout the organic distribution and processing chain. The main objective of such a certification system is to assure consumers that products genuinely have been produced organically. As there is no obvious way for consumers to distinguish whether a product is organic, a price premium for organic products can only be achieved if confidence in the organic quality is ensured. This makes clear standards, certification and labeling of products important factors for the economic success of the organic farming [1].

Standards and Regulations

Essentially the goal of organic standards is to define "organic." Standards should be based on a strict and precise definition of what is organic. They should also be uniform and easy for consumers to understand. Finally standards should bring economic benefit to organic farmers so that their activities are commercially sustainable. Standards serve as a basic guide for organic producers in designing, managing and operating their farms/plantations or processing units.

There are no regulations on organic products that are applicable worldwide. Different associations, industries or governments have varying perceptions of how organic products should be defined and certified. Thus, individual brands naming products organic may in fact have widely different standards for production and certification behind them, giving rise to unfair competitive advantages inthe trade of organic agricultural products. However, work at the international level has progressed towards a common understanding of what organic agricultural products means. One pertains to the development of standards for organic production and trade in the FAO/WHO Codex Alimentarius Commission. The other forum where standards are being developed at the international level is IFOAM.

IFOAM

IFOAM (International Federation of Organic Agriculture Movement) was established in 1972 as an umbrella organization for national organic agriculture associations, but members also include certification bodies, traders and processors. It has established international basic standards (IBS) of organic agriculture and food processing. The standards are updated regularly by the IFOAM Standards Committee and are approved by the IFOAM General Assembly every second year. It has consultative status with the EU and the Codex Alimentarius Commission and a formal liaison with FAO. IBS define how organic products are grown, produced, processed and handled. It provides a framework for certification bodies and standard-setting organizations worldwide to develop their own certification standards. Producer and processors that sell organic products are expected to work within, and be certified by certification bodies, using standards that meet or exceed the requirements of the IBS. This requires a system of regular inspection and certification designed to ensure the credibility of organically certified products and build consumer trust [2].

At the homepage of IFOAM http://www.ifoam.org under "Organic Guarantee System" the IFOAM Norms, consisting of the IFOAM basic standards for organic production and processing and the IFOAM accreditation criteria for bodies certifying organic production and processing are published.

Codex Alimentarius Commission

The Commission is an intergovernmental body that was created in order to protect consumers from health hazards and to facilitate

Box 12.1

Benefits and Threats of Organic Farming Standards and Regulations

Benefits of standards:

- They help organic agriculture to gain a higher profile by giving clear definitions, terms and concepts.
- Standards create consumer confidence and protect consumers from fraudulent products.
- Standards can create innovations and push scientific research through their demand for "organic solutions" in various fields.
- The harmonization and equivalence inherent in standards help organic production to become mainstream with mutual recognition, facilitating market access to thousands of farmers in the market.

Threats posed by organic farming standards:

- Instead of consensus, standards create new borders and trade barriers.
- Equivalence and homogenization standards kill the possibility of local identification and local adaptation.
- Northern standards and procedures can "overrule" sound local agricultural practices and thereby impair indigenous knowledge and development. Overregulation demoralizes farmers' initiative; and it makes innovative development and adaptations to local conditions almost impossible.
- Regulations bring in high costs of certification and market access. Certification keeps farmers occupied with the administration of a mandatory system and becomes a time and energy-consuming burden. The inspection and certification regime does not conform to the cultural context of many of the farmers, i.e., the way they see themselves. This can lead to a total rejection of the whole idea of organic farming.
- The recent developments in setting and executing standards by governments (and no longer by involved NGOs) make the government responsible for defining and setting standards in organic farming and leave the organic farmers' movement worldwide with little or no power.
- When regulations increase, the tendency to cheat might increase.

Source: Ref. 3.

international trade in food products. Two Codex committees are currently developing standards that are relevant to organic products. The formulation of these guidelines is largely based on EU regulations on organic food products and IFOAM standards, as will be discussed below. These guidelines can be very useful to countries that are now drafting their country specific regulations on organic food products.

These Codex guidelines apply to: 1. plants and plant products; 2. livestock and livestock products and 3. handling, storage, transportation, processing and packaging [4]. These Codex Guidelines for organically produced food have been regularly reviewed at least every four years based on given Codex procedure[2]. Further information about Codex Alimentarius is available via the homepage www.codexalimentarius.net. There is also a special homepage on organic agriculture at the FAO Homepage:www.fao.org/organicag/. The Codex Alimentarius-Guidelines on organic agriculture can be down loaded at ftp://ftp.fao.org/codex/ standard/en/ CXG_032e.pdf.

Box 12.2

Content of Organic Farming European Council Regulation 2092/91

- Protects producers from unfair competition;
- Protects consumers from pseudo-organic;
- Protects labels such as "organic," and "biologic";
- Sets rules for production, processing and trade;
- Sets rules for monitoring and certification;
- Sets rules for import.

Source: Ref. 3.

European Union Regulation

Council Regulation (EEC) 2092/91 of 24 June 1991 and subsequent amendments establishes uniform standards and the main rules governing organic production, importation, labeling and marketing of organic agricultural products. The use of the term "organic" is now limited to products that conform to the Regulation. Non-EU products need to show that certification has been carried out by a national authority (in the exporting country) whose equivalence has been approved according to Regulation 2092/91 or by an EU inspectorate which operates internationally like Ecocert, SKAL, KRAV etc. Once a product enters one EU country, it should be able to move across national borders without further inspections. Even then each EU country has its own national standards that in many cases are more stringent than the EU level standards. EU standards should therefore be considered minimum level standards.

The IFOAM EU Group has recently published its position paper on the revision of the organic regulation, EC Regulation 2092/91.

The Group proposed a thorough overhaul of the regulation, starting from basic principles that lead to certifiable standards. It was opined that this regulation has been of great importance to the growth of the organic sector and time to ensure the standards better reflect the many changes of the past and can cope with the dynamics and challenges of the future. The European Commission aims to present a proposal for the revision of the organic regulation to the Council by end of the year, 2005. The Council will deal with the Commission proposal under the coming Austrian presidency (Source: www.ifoam.org).

The EUR-Lex website leads to a consolidated version of the EU Regulation 2092/91 and includes amendments up to 23.03.2002. Available in all languages of the EU at http:// europa.eu.int/eurlex/ de/consleg/main/1991/en_1991R2092_in dex.html

EU Logo for Organic Products

In February 2000 the European Commission introduced a logo for organic products that may be used throughout the EU by producers operating in accordance with the provisions of the regulation on organic production. Consumers buying products bearing this logo can be confident that:

1. at least 95 per cent of the product's ingredients have been organically produced;
2. the product complies with the rules of the official inspection scheme;
3. the product has come directly from the producer or preparer in a sealed package;
4. the product bears the name of the producer, the preparer or vendor and the name or code of the inspection body.

The use of the symbol is voluntary, and it may also be used in conjunction with national government or private logos for identifying organic products. So far only few companies, especially in Southern Europe, are using the EU logo (Fig. 12.1) and the market impact are low. But EU logo has some advantages and drawbacks [1].

Fig. 12.1. EU logos for designated origin compared to the organic farming logo.

The advantages are:

- It is free of charge. Anybody who produces certified organic products may use it at no additional cost.
- It is applicable in all EU countries, thus facilitating trade.

The drawbacks are:

- It is neither actively promoted nor supported by a marketing campaign.
- It is very similar to other European logos under the 'designated origin' legislation (Council Regulation 2081/92 and 2082/92 (Fig. 12.2).
- It cannot be used for organic products with more than 5 percent imported (to the EU) ingredients. This is a major obstacle to the use of the logo in many cases.

In 1992, the European Union created systems known as PDO (Protected Designation of Origin), PGI (Protected Geographical Indication) and TSG (Traditional Speciality Guaranteed) to promote and protect food products.

Other National Regulations

Many countries outside the European Union legally protect organic products or are in the process of development of organic regulations. All these regulations laydown minimum rules governing the production, processing and import of organic products, including inspection procedures, labelling and marketing[2].

Several EU countries have developed their own national regulations as well as national logos for organic products; in some cases this occurred long before the EU regulation on organic production came into force. These logos are well known and much trusted by consumers. The existence of these logos is one reason for the organic boom in these countries (Table 12.1)

United States Organic Standards

In the United States, the Organic Foods Production Act (OFPA) of 1990 required the United States Department of Agriculture (USDA) to develop national standards for organically-produced agricultural products and to establish an organic certification programme, based on recommendations of the National Organic Standards Board (NOSB). The National Organic Program (NOP) (may see http://www.ams.usda. gov/nop/ ConsumersLinks.html) is a marketing programme housed within the USDA Agricultural Marketing Service

Table 12.1. Government and private logos for organic products in Europe

Austria (state)	Belgium (private)	Czech Republic (state)
Denmark (state)	Finland (state)	France (state)
Germany (state)	The Netherlands (state)	Norway (private)
Spain (state)	Sweden (private)	Switzerland (private)

(may see http://www.ams.usda.gov/), the agency that sets marketing standards, first proposed draft standards in December 1997 and a revised proposal was issued in March 2000. The final regulation, adopted in December 2000, was fully implemented in October 2002[4]. The primary purpose of the NOS programme is to standardise organic production standards in the US and to prevent fraud and mislabeling of products. All organic foods marketed in the US are produced according to national standards will bear the official USDA Organic seal. Further information is available at the site of the American Marketing Services (http: //www.ams.usda.gov).

Standards have been developed, based on:

- NOSB recommendations;
- A USDA review of state, private and foreign organic certification programmes;
- Public consultations. The fact that USDA received over 300,000 public comments demonstrates the great interest in the development of standards for organic products.

The USDA has introduced four types of organic labels (http:// www.organicmonitor.com/ r2110.htm):

- 100 percent organic: Only organically-produced ingredients are in the product.
- Organic: Contains 95 to 99 percent organically produced ingredients. Food may contain spices or flavorings that are non-organic.
- Made with organic ingredients: At least 70 percent organic ingredients.
- Some organic ingredients: Below 70 percent.

The organic seal can be used on the first two products that are 100 percent organic and organic; however there would be no seal for the latter two products. All organic foods in the US must meet national standards of production and companies face a fine if their organic products do not meet these standards. This also applies to imported organic products.

Several comparisons have been made between the different standards and guidelines, such as between IFOAM Basic Standards, Codex Alimentarius Guidelines for Organically Produced Food, and EU Regulation 2092/9[5], as well as between the EU and the United States[6]. There are important differences between the USDA and EU standards. As explained on the USDA website, "USDA national organic standards should be considered in two segments: (1) the verification system that includes certification of organic agricultural products and accreditation of certifying agents and (2) the production, handling and labelling standards under which organic agricultural products are produced and sold. NOP is consistent with the internationally accepted guidelines for certification and accreditation, International Standardization Organization Guides 65 and 61 respectively [4]. However, NOP requirements for production, handling

Box 12.3

Summary of Provisions of Some Standards

USA - National Organic Programme

- Certification by USDA accredited agents is mandatory.
- US government enforcement with civil penalty for violations.
- 36 months land transition - no exceptions.
- No synthetic pesticides or fertilizers.
- No GMO's or their products.
- Organic seed required.
- Annual farm plan.
- National materials list.

Europe - EEC 2092/91 and Beyond

- Foundation standard is EU Council Regulation 2092/91, published in 1991.
- In each member state a 'competent authority' interprets the basic standard and accredits certifiers.
- Many certifiers in Europe have established standards that are more restrictive than the basic standard.
- European certifiers often do not recognize each other.
- Products from outside EU (and exempt 'second countries') must receive an import authorization from the competent authority.
- The EU has initiated, in early 2002, bilateral negotiations with the USDA for equivalency.

Japan Agricultural Standards

- Implemented by Japan Ministry of Agriculture, Forestry & Fisheries (MAFF) in April 2001.
- Product sold as organic must bear the JAS seal.
- In March 2002, MAFF recognized the US National Organic Programme as equivalent for plant-based ingredients.
- JAS mark must be affixed by JAS certified entity.

and labelling, and allowed and prohibited materials differ significantly from those of other countries such as those of the EU, particularly in terms of livestock production standards."

Discussions between the EU and U.S. have been ongoing with regard to equivalency between the organic standards of the two regions. But they have not yet reached an agrrement about the equivalency. However, the U.S has recognition agreements with two European nations, the UK and Denmark, in which accredited agents of these two governments can certify that products were produced or handled in accordance with USDA regulations.

The Japan Agricultural Standards (JAS) for organic foods are in some respects, less stringent than those in the US. Products are not evaluated on an "organic/non-organic" scale like in the US, but are classified by the degree of organic farming employed. While the JAS certification lends official credibility to claims of low-chemical farming methods, exporters should be aware that under current Japanese law, manufacturers that do not obtain the JAS seal and yet might still label their products as "organic". JAS is a voluntary system rather than a truth in advertising law. However, the benefits of having the Japanese government's official seal backing up an imported food product's safety and healthiness are obvious in making the product more attractive to consumers and the Japanese trade.

Demeter -International eV

Demeter International eV is a worldwide network of 19 international certification bodies. It claims to have more than 3,500 partners in 35 countries, covering 1 million hectares of cultivated land. Its internationally recognized mark is registered in over 60 countries. Further information is provided at its website (http:www.demeter.net).

The customer has implicit trust in Demeter products, and relies on the fact that the Biodynamic quality is guaranteed from the field to the consumer. All enterprises are inspected at least once a year by qualified, independent inspectors from an accredited inspection agency. As well as the requirements contained in EU 2092/91, and other organic standards, the audit covers the additional demands of the Demeter standards.

The inspection report is the basis for a decision, made by an expert committee in favour of, or against, certification. The use of the Demeter trademark is governed by a contract with the trademark owner. The conversion usually takes three years to achieve Demeter certification. Certification as "In conversion to Demeter" can be attained after farming the whole enterprise biodynamically according to the International Demeter Standards for at least 12 months. If it can be proven that previous farming methods were organic or extensive, the time to full certification can be shortened. In comparison to the standards for organic agriculture, some additional requirements must be met, such as:

- the conversion usually takes three years to achieve Demeter certification;

- cattle or other ruminants must be kept on agricultural farms;
- soil fertility is to be maintained primarily through the use of well rotted compost, which has been prepared with the Biodynamic compost preparations;
- all areas are to be sprayed with the Biodynamic horn dung and horn silica preparations;
- no use of copper in vegetables;
- use of copper on perennial crops is restricted to maximum 3kg/ ha/year averaged over 5 years.

Co-operation with an internationally experienced advisor is essential for conscientious conversion planning. Questions concerning fertilisers, plant protection and crop rotations, or about the production and use of the biodynamic® preparations can be discussed in depth. The subsequent process is to educate local people so that in time they can take over and provide support for the farmers. Further information is provided at its website (http://www.demeter.net).

Certification and Accreditation

Organic agriculture has special needs for production planning and management beyond the traditional farming because of the limitation imposed by the terms of organic registration and certification [7]. The organic guarantee system is a way of assuring consumers that the foods they are purchasing are organically produced. Broadly, this is done in two ways: one way is the producer's personal guarantee (that does not involve certification), and the other is third-party certification, a system by which conformity to applicable standards is determined and confirmed by a third party or an independent body [4]. IFOAM provides framework of certification programmes worldwide and encourages to develop own national or regional standards. These will take into account local conditions resources and may well be more detailed than the IFOAM Basic Standards (IBS). Many private certification bodies have developed their own certification programmes, generally based on IBS, and they use their own trademarks. Governments have played an increasingly important role in developing organic standards and accreditation programmes. The Organic Certification Directory 2003, published by Grolink, lists 364 bodies as offering organic certification services. Out of 364 certification bodies, 290 of them are located in the European Union (106 certifying bodies), the USA, Japan, Canada and Brazil (184 certifying bodies). 56 of the listed certification organizations also operate outside their home country. Most of them are based in a developed country and offer their certification services in developing countries. There are only 57 countries that have a home-based certification organization. Most of Africa and Asia are still lacking local service providers. There are only 7 certification bodies in Africa and most of them are in South Africa. Out of 83 in Asia, 65 are in Japan [8].

Farmers and marketing farms seeking to sell their products in developed countries must usually apply to and hire an organic certification agency for inspection and confirmation that they adhere to the standards established by various trading partners. The guarantee system also includes accreditation of certifiers. National accreditation is sufficient for the domestic market but international accreditation is necessary for certifiers whose clients want to access foreign markets. In addition, there are regulations that have to be followed demanded in the countries of operation as well as in the countries to export.

Certification of organic production is generally realized in three steps.

- *Producer*: The producers and the fields and facilities used in the production.
- *Production system*: Production, handling and processing methods.
- *Products*: Finally, products are labeled.

The basic requirements for certification for organic products can be summarized[9] as follows:

- minimum one full inspection per year;
- full implementation of the relevant standards;
- conversion period to be followed;
- no parallel production;
- sustainable production system;
- identification of product flow and audit procedure; and
- clear management responsibilities.

An organic certification system normally has the following four elements:

Standards: There must be standards, which are clearly formulated and communicated among all parties of the certification system as well as the parties concerned.

Contracts and legal framework: All producers within a certification system should be bound by written agreement. The whole system must be handled by a body that has legal status for registration and certification.

Inspection: It must cover production, transactions, storage, processing, labeling and certification. The certification body taps the services of independent inspectors. The inspector looks into the farm design and whether and how the farmer utilizes the desired processes in organic production. The farm/planatation or production system is certified as organic or not, based on the inspector's report and the applicant's compliance with other requirements of the certification programme. Inspection is a task that requires technical knowledge and practical experience, preferably in both organic and conventional production systems. In addition, inspectors must have communication skills, investigative skills, and good working relations with producers. Moreover, they must be imaginative and resourceful, and their integrity and credibility must be well recognized. It is the inspector's report that serves as the main basis for certification by the committee. The inspection process has its own manual of operations and inspection programmes. In Asia trained inspectors are only found in countries that have certifying agencies.

Certification: The documentation pertaining to the inspection is given to the certification body that will then usually have a certification committee whose task is to assign the appropriate status to the information contained in the application for certification in combination with the inspection report.

Box 12.4

Certification Process

The following is a simplified description of the route to organic certification.

1. Choose a certification body that meets the need.
2. Apply for and fill in all requested documents. The same will have to be returned for review of the certification authority.
3. An inspection assignment is drawn up contracting the inspector for a site visit.
4. An inspection costing (quote) is drawn up by the inspector and submitted to the applicant.
5. Agreement between the parties is reached.
6. The actual inspection/verification takes place.
7. A certification decision is reached by the certification committee based on the documentation prepared during inspection.
8. A contract is then entered into between the applicant and the certification body with regard to the status achieved.

Possible outcomes:

- Full organic status
- Full organic status with condition
- Organic in conversion
- Organic in conversion with conditions
- Denial for certification

There are some important points related with inspection and certification of a tea estate, which are to be kept in mind.

Soil analysis

Soil analyses of zero to 30 cm depth should be necessarily done each year, regularly, for macro and micronutrients and heavy metal status. This will reveal fertilizer usage and contamination from sources other than intentional application such as wind drift from neighboring plantations or water migrating from plantations above.

Leaf analysis

Tissue analysis of mother leaf should also be carried out, annually, for pesticide and fungicide residues and for the levels of macro and micronutrients and heavy metals. This will reveal mineral nutrient status and pesticide and fungicide usage, if any.

Marketable tea analysis

The marketable organic tea should be analysed twice a year, pre- and post- monsoon, for pesticide residues, flavour and quality.

Book-keeping

Book-keeping should be properly maintained to make an audit trail of inputs and marketable tea supplied to consumer. The invoice number should be mentioned on tea chests, in the factory book and in the documents. With the invoice number, the origin of the tea can be traced back to when it was manufactured, the quantity manufactured and the site and quantity of leaves plucked in the field with the factory diary and field book.

Inspection by independent agency

Surprise inspection by the representatives of an independent agency unannounced at least once a year is required by field visits to check the organic growing techniques, taking samples of leaf and marketable tea at random and also to check the maintenance of book-keeping system.

Certification is performed for a fee and those farms not meeting the standards of accrediting agencies pay the fee even when they fail, although failure does not mean a complete loss of opportunity to sell as organic. Farmers are given suggestions for upgrading to the level of certification, and the certification process can be performed again the next season. All organic tea farms should have annual certifications.

The high cost of certification (may see Table 1.2) is, however, a major impediment to increasing certified agricultural production in developing countries. Box I2.5 discusses several steps that could be taken to reduce such costs. Another important issue for developing countries is how to make certification more affordable for small producers. It may, therefore, be useful to explore possibilities to facilitate certification of products through mechanisms such as "umbrella certification" of certain products (i.e. certification of entire geographical areas or groups of producers rather than individual enterprises), or the development of regional certification bodies. IFOAM Basic Standards have provisions for group certification.

Box 12.5

Reducing Certification Costs in Developing Countries

This could be achieved by the following means:

1. Assistance from donors, as well as cost-sharing with developed country partners, for example within the framework of fair-trade and development projects. This may be the preferred option of small countries, in particular LDCs (least developed countries), as well as small rural and indigenous communities in both LDCs and other developing countries.
2. Training of local inspectors and other personnel who work for an international certification body operation in a particular developing country. This could involve a form of co-certification. For example, Demeter and SKAL International, the largest international certification bodies operating in India, employ local personnel.
3. Development of a national certification system, with the assistance of IFOAM or bilateral/multilateral donors, as well as Government support. Developing countries with a relatively larger organic sector and some certification infrastructure, such as Brazil, Costa Rica, India and South Africa, have taken steps in this direction. Brazil already has a well-developed certification infrastructure. Brazil have certification bodies accredited by IFOAM's International Organic Accreditation Service (IOAS), and by other competent authorities in major importing countries.
4. Negotiation of equivalency agreements, such as, recognition under European Union Regulation 2092/91. India, for example, has taken steps in this direction.
5. Regional cooperation.

Source: adapted from UNCTAD/COM/88, pp. 31-33

The use of different logos in the importing countries is an additional problem for the export of organic products from developing countries. Though the EU Directive and Council Regulation 2092/2091 on organic products cover all member States, national organic schemes and labels continue to differ from each other. Policies on organic food procurement also tend to differ, as does consumer choice. In markets such as Belgium, Denmark, the Netherlands, and Sweden, certified organic quality is conveyed to the consumer by a single logo. In France, before the introduction of the Agriculture Biologique (AB) logo, several different logos dominated the market. In Germany, a single logo was introduced in 1999, which can be used for the whole country. All these are expensive and cumbersome to developing country suppliers [4].

Accreditation

Certification is a market instrument. The certification services are being rendered worldwide by a number of certifying bodies. However, certifying bodies are subject to scrutiny by national or international organizations to ensure an acceptable level of services. If these are satisfied with the performance, accreditation follows. In order to assure the public's trust in the certification organizations, accrediting organizations should be perceived as guardians of public health, food safety and sound environment. They should not be perceived as a higher level of bureaucratic regulation or as a protectionist instrument [4].

In 1992, IFOAM established the IFOAM-Accreditation Program (managed by the International Organic Accreditation Service, IOAS [10]) to support international equivalency for organic quality claims. In order to harmonize accreditation according to the principles of the IFOAM-accreditation program, the IFOAM is currently negotiating with governments that have set standards for organic farming and with authorities that give import permits. However, accreditation is still a playground for false competition between private and government accreditors [11]. Extensive information about the IFOAM accreditation programme is available at the IOAS Home page at http://www.ioas.org/.

The International Organic Accreditation Service (IOAS)

The IOAS is a non-profit, independent company registered and headquartered in the United States with satellite offices in Europe. It was established by IFOAM in 1997 to run the IFOAM Accreditation

Programme, which had been run by IFOAM itself since 1992. The IOAS operates the Accreditation Programme under licence from IFOAM. The IOAS are working towards a collaboration with goverments and international organizations both multilaterally within the International Task Force on Harmonization set up by FAO, IFOAM

Box 12.6

Quick Overview of the IFOAM Accreditation Application Process

The certification body (CB) completes filling up of application form, collates necessary documentation and completes document checklist. Duly filled in application along with all required documents and application fee should be sent.

Once the application is received the IOAS conducts detailed scrutiny of the documentation and prepares the screening report and informs applicant of the non-compliances if any found in the screening of documentation. These are noted as nonconformities, deficiencies and observations. CB is invited to supply evidence of corrective actions to remedy all nonconformities within 3 months. A copy of the screening report will be supplied to the CB. The IOAS reviews the corrective actions and if these are satisfactory an evaluation visit is organised. If they are not satisfactory the IOAS may allow an additional period for compliance. An evaluation visit contract is sent. Contract includes name(s) of evaluator(s) and cost. An estimate of the evaluation costs is made and an invoice for 70 percent of these is sent to CB. This must be paid prior to the visit.

The evaluator will arrange his visit with the applicant CB. The visit will be made and a report compiled. The IOAS will review the report and inform the CB of any additional nonconformities or deficiencies. The CB will be required to correct all nonconformities within three months for accreditation to be possible. A copy of the visit report will be sent to the CB. The remaining 30 percent of the evaluation fee together with the assessment fee is paid.

The IOAS Accreditation Committee reviews the corrective actions and if these are satisfactory a contract will be offered. The accreditation contract includes any conditions of accreditation (corrective actions following accreditation) with an agreed timeline depending on the nature of the nonconformity. The contract is signed and returned to the IOAS.

During the process, the CB may appeal overall decisions and may also challenge the justification for individual nonconformities. The appropriate portion of the annual fee is paid on signing the contract. A Certificate of Accreditation will be issued to the accredited organisation together with a copy of the signed contract. At least one year would be required to complete the application process.

Source: http://www.ioas.org/WEBSITE/0410_IFOAM_Accreditationfinal.htm

and UNCTAD and bilaterally to look at ways to share information and reduce the regulatory burden on certification bodies and in turn producers.

The sole purpose of the IOAS is to offer international oversight of organic certification through an accreditation process for certification bodies active in this field.

The standards and requirements against which applicants are judged are the IFOAM Basic Standards and Criteria – requirements developed by the international organic industry with organic integrity, transparency and stakeholder participation as core values.

The IFOAM Accreditation Programme is an industry-based global guarantee of organic integrity, unburdened by national barriers and implemented by one body which has no other interests.

Through a rigorous audit process that complies with international accreditation norms, IOAS help in building a common organic currency, so that producers, processers and traders can go about their business confidently, and consumers can trust the label they read.

Certifiers that have been accredited by IFOAM have developed and signed an agreement to recognize the certifications of other IFOAM-accredited certifiers. The agreement establishes a framework for mutual recognition based on IFOAM accreditation. This agreement has been used as the basis for certification transference.

In addition to IFOAM accreditation, the IOAS also offers accreditation against ISO/IEC Guide 65 General Requirements for bodies operating product certification system and compiles reports addressing a certification body's conformity with organic regulations such as EU Regulation 2092/91.

Codex Alimentarius

According to the Codex Alimentarius Guidelines, import requirements should be based on the principles of transparency and equivalency as set out in the Principles for Food Import and Export Inspection and Certification. In accepting imports of organic products, countries would usually assess the inspection and certification procedures and the standards applied in the exporting country. Imported products may be marketed as "organic" only where the competent authority or designated body in the exporting country has issued a certificate of inspection. According to Article 7.4, an importing country [4] may:

- require detailed information, including reports established by independent experts;
- arrange, together with the exporting country, for site visits to examine inspection, certification and production in the exporting country; and
- require to avoiding consumer confusion, that the product is labelled in accordance with the labelling requirements of the importing country.

European Union

There are three ways to export organic products to the EU which are based on Article 11 of Regulation 2092/91 EEC [4]:

- paragraph 1: Establishes a 'third-country' list, indicating countries with which equivalence is established. However, only eight countries are on the list: Argentina, Australia, Costa Rica, the Czech Republic, Israel, Hungary, New Zealand and Switzerland;
- paragraph 6: Organic products from countries which are not on the "third-country" list can be marketed in the EU, provided the importer submits documentation to confirm that the products are produced and certified according to rules equivalent to those of EU. Such authorization shall be valid only as long as these conditions are shown to be satisfied. Over 70 developing countries export to the EU under this provision.
- paragraph 7: An EU member state assesses an inspection body in a third country and requests the commission to approve it. The commission can then add that inspection body to the Article 11 list.

United States

Accreditation of goods to be imported into the United States [4] from foreign countries can occur in three ways:

- certifying agents operating in foreign countries may apply for USDA accreditation;
- the USDA determines, upon the request of a foreign government, that that country's authorities are able to assess and accredit certifying agents as meeting the requirements of the NOP;

- the USDA and a foreign government agree upon equivalency of standards and certification procedures, so that organic imports from this country are acceptable in the United States.

Japan

MAFF issued organic regulations in early 2000 (Notifications No.59 and 60). They provide three options for imports of organic products into Japan [4]:

- certification by a MAFF-registered certification organization (RCO) in Japan, where the RCO in Japan certifies the production/ processing in the exporting country;
- certification by a MAFF registered foreign certification organization (RFCO) in the exporting country;
- re-certification, carried out by an RCO in Japan, and needed for imported goods destined for use as ingredients in finished products marketed as organic in Japan.

International certification organizations from the USA and Europe have been working as the accreditation body in Latin America, Asia and Africa. With some exceptions there are very few local certification bodies established in developing countries. It is a big task to establish local certification bodies as long as they are aiming at export. Furthermore, international nonprofit organizations such as FAO and ITC are devoting resources to assess and announce opportunities for developing countries to supply organic products to the major consumer markets [12].

The inspection itself can be very cumbersome. In most Western countries there is an idea that all fields should be inspected by the external certification bodies. If we consider the availability of a number of the parcels and the fact that they are so scattered, in such cases inspection of all fields increases the costs greatly. Therefore, internal control by a local organization, evaluation of the internal control system, and random inspection by external certification could be advised [13].

Furthermore, the standards and certification systems for organic agriculture have been developed without sufficient participation from developing countries, and do not reflect particular needs or circumstances of these countries i.e. their traditions, cultures and existing infrastructure.

IFOAM Accredited Certification Bodies

1. **Australian Certified Organic (ACO)**

   P. O. Box 3404, Toowoomba Village Fair, Queensland, 4350, Australia tel. +61 7 4639 3299, fax +61 7 4639 3755, e-mail: manager@bfa.com.au

   Countries of Operation: Australia, Fiji, Hong Kong, Japan, New Zealand, Papua New Guinea, Singapore.

2. **Organic Agriculture Certification Thailand (ACT)**

   619/43 Kiatngamwong Building (1st floor), Ngamwongwan Road, Tambon Bangkhen Muaung, Nonthaburi province, 11000, Thailand tel. +66 2 952 6677, fax + 66 2 580 0934, e-mail: actnet@ksc.th.com Countries of Operation: Thailand.

3. **AGRIOR Ltd.**

   121 Hachashmona'im St., Tel Aviv, 67011, Israel

   tel. +972 3 5614898, fax +972 3 6241897, e-mail: agrior@net vision.net.il

   Countries of Operation: Israel, Ethiopia.

4. **AgriQuality Ltd.**

   PO Box 4127, Mount Maunganui South, Hamilton, New Zealand tel. + 64 7 572 0814, fax +64 7 572 0839, e-mail: schofieldh@certenz.co.nz

   Countries of Operation: New Zealand, Vanuatu.

5. **ARGENCERT S.R.L**

   Bernado de Irigoyen 760, 10"B, 1072 Buenos Aires, Argentina tel. + 54 11 4342 1479, fax +54 11 4 331 7185, e-mail: argencert@argencert.com.ar

   Countries of Operation: Argentina, Chile, Paraguay.

6. **Bioagricert srl**

   Via del Macabraccia 8, 40033 Casalecchio di Reno (BO), Italy tel. + 39 051 562 158, fax +39 051 562 294, e-mail: riccardo@bioagricoop.it

   Countries of Operation: Italy, Bulgaria, Colombia, Ecuador, Mexico, Thailand, Tunisia, Turkey, Vietnam.

7. **BIO-GRO New Zealand**

   PO Box 9693 Marion Square, W, Wellington 6031, New Zealand

tel. + 64 4 801 9741, fax +64 4 801 9742, e-mail: smason@bio-gro.co.nz

Countries of Operation: New Zealand, Fiji, Cook Islands, Niue, South Africa.

8. **Bioland e. V.**

Kaiserstrasse 18, D-55116 Mainz, Germany tel. +49 61312397924, fax + 49 613123979-27, e-mail: landbau@bioland.de

Countries of Operation: Germany, Belgium, France, Italy, The Netherlands.

9. **BIOPARK e.V.**

Karl-Liebknecht Strasse 26, D-19395 Karow, Germany tel. + 4938738-70309, fax + 4938738-70024, e-mail: info@biopark.de

Countries of Operation: Germany.

10. **BOLICERT**

Casilla 13030, General Gonzálves, 1317, La Paz, Bolovia tel. + 591 2 490747, fax + 591 2 490747, e-mail:bolicert@mail.megalink.com

Countries of Operation: Bolivia, Paraguay.

11. **California Certified Oganic Farmers**

1115 Mission Street, Santa Cruz, CA 95060, USA tel. +1 831 423 2263, fax + 1 831 423 4528, e-mail: Brian@ccof.org

Countries of Operation: USA, Canada, Mexico.

12. **Consorzio per il Controllo dei Prodotti Biologici (CCPB)**

Via Jacopo Barozzi N.8, 40126 Bologna, Italy tel. +39 0 51 6089811, fax + 39 0 51 254842, e-mail: ccpb@ccpb.it

Countries of Operation: Italy.

13. **Ekoagros**

K. Donelaicio str. 33, LT - 3000, Kaunas, Lithuania tel. + 370 37 20 31 81, fax + 370 37 20 31 82, e-mail: ekoagros@ekoagros.lt

Countries of Operation: Lithuania.

14. **INSTITUTO BIODINAMICO**

Caixa Postal 321, CEP 18603-970, Botucatu SP, Bra zil tel. + 55 14 6822 5066, fax +55 14 6822 5066, e-mail: ibd@ibd.com.br

Countries of Operation: Brazil, Bolivia, Dominican Republic, Paraguay.

15. **International Certification Services Inc.**

    301 5th Ave. SE, Medina, ND 58467, USA tel. +1 701 486 3578, fax +1 701 486 3580, e-mail: Info@ics-intl.com

    Countries of Operation: USA, Brazil, Canada, Mexico, Nepal, Paraguay.

16. **Istituto Meditera neo Di Certificazione s. r. l. (IMC)**

    Via Carlo Pisacane, 32, 60019 Senigallia, Ancona, Italy tel. + 39 71 792 8725, fax + 39 71 791 0043, e-mail: imcert@tin.it

    Countries of Operation: Italy, Tunisia, Egypt, Philippines.

17. **Istituto per la Certificazione Etica e Ambientale (ICEA)**

    Stra da Maggiore 29, 40125, Bologna, Italy tel. + 39 0 51 272986, fax + 39 0 51 232011, e-mail: icea@icea.info

    Countries of Operation: Italy, Bosnia- Herzogovina, Kazakhstan, Romania, Ukraine, Uruguay.

18. **Japan Organic & Natural Foods Association**

    Takegashi Bldg., 3F, 3-5-3, Kyobashi, Chuo-Ku, Tokyo, 104-0031, Japan tel. +81 3 3538 1851, fax +81 3 3538 1852, e-mail: matsumoto@jona-japan.org

    Countries of Operation: Japan, China.

19. **KEZ**

    Podébradova 909, 537 01 Chrudim, Czech Republic tel. + 420 455622249, fax + 420 455622249, e-mail: tomaszidek@quick.cz

    Countries of Operation: Czech Republic.

20. **KRAV-Ekonomisk Förening**

    Box 1940, S-751 49, Uppsala, Sweden

    tel. +46 181 00290, fax + 46 181 00366, e-mail: johan.cejie@krav.se

    Countries of Operation: Bosnia Herzogovina, Denmark, Finland, Malaysia, PR China, Peru, Russia, Spain, Sweden, Tanzania, Uganda.

21. **National Association Sustainable Agriculture Australia (NASAA)**

    PO Box 768, Stirling 5152, South Australia, Austra lia tel. + 61 88 3708455, fax + 61 88 3708381

    e-mail: admin.manager@nasaa.com.au

    Countries of Operation: Australia, East Timor, Fiji, Indonesia, Japan, Malaysia, Nepal, New Zealand, Papua New Guinea, Samoa, Sri Lanka.

22. **NATURLAND e.V.**

Kleinhaderner Weg 1, 82166 Gräfelfing, Germany tel. + 49 89 8980820, fax +49 8989 808290, e-mail: naturland@naturland.de

Countries of Operation: Germany, Argentina, Austria, Belgium, Bolivia, Brazil, Chile, Colombia, Croatia, Dominican Republic, Ecuador, Egypt, El Salvador, Greece, Guatemala, Hungary, Indonesia, India, Ireland, Italy, Mexico, The Netherlands, Nicaragua, Paraguay, Peru, Philippines, Spain, Sri Lanka, Switzerland, Uganda, the United Kingdom, Vietnam.

23. **Organic Crop Improvement Association (OCIA)**

6400 Cornhus ker, Suite 125, Lincoln NE 68507, USA tel. + 1 402 477 2323, fax + 1 402 477 4325, e-mail: info@ocia.org

Countries of Operation: USA, Brazil, Canada, China, Columbia, Costa Rica, Dominican Republic, East Timor, Ecuador, Germany, Guatemala, Honduras, Italy, Japan, Mexico, Paraguay, Peru, Philippines, Uganda.

24. **Organic Farmers & Growers LTD**

Elim Centre, Lancaster Road, Shrewsbury, Shropshire, SY1 3LE, United Kingdom tel + 44 1743 440512, fax + 44 1743 461441, e-mail: richard.jacobs@organicfarmers.uk.com

Countries of Operation: United Kingdom, France.

25. **Organic Food Development & Certification Center of China**

8 Jiangwangmiao Street, P.O. Box 4202, Nanjing, 210042, P.R.China tel. + 86 25 5425370, fax + 86 25 5420606, e-mail: ofdcsepa@pub lic1.ptt.js.cn

Countries of Operation: P.R.China.

26. **Organizacion Internacional Agropecuria**

AV. Santa Fe 830, 1641 Acassuso, Buenos Aires, Argentina tel. + 54 11 4793 4340, fax + 54 11 4793 4340, e-mail: oia@oia.com.ar

Countries of Operation: Argentina, Brazil.

27. **Quality Assurance International**

12526 High Bluff Drive, Suite 300, San Diego, CA, 92130, USA tel. +1 858 792 3531, fax + 1 858 792 8665, e-mail: marian@qai-inc.com

Countries of Operation: USA.

28. **Soil Association Certification Ltd.**

    Bristol House, 40-56, Victoria Street, Bristol BS 1 6BY, United Kingdom tel. + 44 117 987 4576, fax + 44 117 925 2504, e-mail: info@soilassociation.org

    Countries of Operation: United Kingdom, Belize, Bosnia, Dominica, Egypt, France, Ghana, Granada, Guyana, Iran, Kenya, Namibia, South Africa, Syria, Thailand, Venezuela, Zambia, Zimbabwe.

IFOAM Accreditation Applicants (9/8/2005)

1. **Debio**

   Bjorkelangen N- 1940, Norway tel: + 47 638 62 650, fax: + 47 638 56 985, e-mail: morten@debio.no

2. **Ecoland e.V.**

   Raiffeisenstraße 18, Schwäbisch Hall 74523, Germany tel: +49 (0) 7919329, fax: +49 (0) 791 93290 4, e-mail: info@ecoland.de

3. **LETIS S.A.**

   Urquiza 1564 (S2000 ANR), Rosario Santa Fe, Argentina tel: +54 341 426 4244, fax: +54 341 426 4244, e-mail: letis@letis.com

Certification Bodies Accredited Against ISO/IEC Guide 65

1. **Agrior Ltd**

   121 Hachashmona'im St., Tel Aviv 67011, Israel tel: +972 3 6241897, fax: + 972 3 6241897, e-mail: agrior@netvision.net.il Scope: IBOAA Private Standards and Agrior Seal Programme.

2. **National Association Sustainable Agriculture Australia**

   PO Box 768, Stirling, S. Australia, 5152, Australia tel: + 61 08 3708381, fax: +61 08 3708381, e-mail: admin.manager@nasaa.com.au Scope: NASAA Private Standards and Seal Programme.

3. **Japan Organic & Natural Foods Association**

   Takegashi Bldg. 3F, 3-5-3, Kyobashi Chuo-Ku, Tokyo, 104-0031 Japan tel: + 81 3 3538 1851, fax: + 81 3 3538 1852, e-mail: matsumoto@jona-japan.org Scope: JONA Private Standards and Seal Programme and Japan Agricultural Standard.

4. **Organic Agriculture Certification Thailand**

   619/43 Kiatngamwong Bldg , Ngamwongwan Rd., Tambon Bangkhen, Muang District, Nonthaburi 11000 Thailand tel: + 66

2 952 6677, fax: + 66 2 580 0934, e-mail: info@actorganic-cert.or.th
Scope: ACT Private Standards and Seal Programme which have been found to be equivalent to EU Regulation 2092/91

5. **Organic Certifiers**

   6500 Casitas Pass Road, Ventura CA 93001 USA tel: + 1 805 684 6494, fax: + 1 805 684 2767, e-mail: organic@west.net

6. **Bolicert**

   Casilla 13030 General Gonzálves 1317 La Paz, Bolivia tel: + 59 12 2490747, fax: + 59 12 2490747, e-mail: bolicert@mail.megalink.com

7. **ETKO**

   160 sok. No 13/7 35 040, Bornova, Izmir, Turkey tel: +90 232 339 76 06, fax: +90 232 339 76 07, e-mail: ma@etko.org

Demeter Certification - Approved Organic Inspection Agencies

BCS Öko-Garantie GmbH

Cimbernstr. 21
D-90402 Nürnberg, Germany
phone: + 49-911-424390, fax: + 49-911-492239
Email: info@bcs-oeko.de
internet: www.bcs-oeko.de
Demeter inspections in: World-wide (Costa Rica, Dom. Rep., Ecuador)

Biokontroll, Hungària KHT.
Margit krt. 1. III/16-17
H-1027 Budapest, Hungaria
phone: + 36-1-336112-2, fax: + 36-1-336112-3
Email: info@biokontroll.hu
internet: www.biokontroll.hu
Demeter inspections in:Hungary

CERES - Certification of Environmental Standards GmbH (Ltd.)
Vorderhaslach 1
D-91230 Happurg, Germany
phone:+ 49-9158-928291, fax: + 49-9158-9289862
Email: schulz@ceres-cert.com
internet: www.ceres-cert.com
Demeter inspections in: Middle- and South America, Eastern Europe.

ECOCERT International
Güterbahnhofstr. 10
D-37154 Northeim, Germany
phone: + 49-5551- 90 84 30, fax:+ 49-5551-90 84 380
Email: info@ecocert.de
Demeter inspections in: World-wide (Morocco, Tunisia)

Hungaria Oeko Garancia GmbH
Inspection and Certification Ltd.
Frankel Leo u. 5 I. em 12/a
H-1027 Budapest, Hungaria
phone: + 36-1-336-053-3 or -4
fax: + 36-1-316-53-30
Email: hoeg@axelero.hu
Demeter inspections in: Hungary, Romania.

IMO (Institute for Marketecology)
Weststr. 51
CH-8570 Weinfelden, Switzerland
phone:+ 41-716-26 0 626, Fax: 41-716-26 0 623
Email: imo@imo.ch
internet: www.imo.ch
Demeter inspections in: World-wide (Chile, Dom. Rep., Greece, Sicily, India, Japan, Mexico, Poland, Sri Lanka, Turkey).

Instituto Biodinamico
C.P. 321
BR-18603-970 Botucatu/Sao Paulo
Brasil
phone: + 55-146-8225066, fax: + 55-146-8225066
Email: ibd@ibd.com.br
internet: www.ibd.com.br
Demeter inspections in: Middle and South America.

Lacon GmbH
Weingartenstr. 15
D-77654 Offenburg, Germany
phone:+ 49-7819-1937-30
fax:+ 49-7819-1937-50
Email: lacon@lacon-institut.com
internet: www.lacon-institut.com
Demeter inspections in: World-wide, prior Africa (Tunisia, Cyprus)

KGZS
Institute of Agriculture
Vinarska 14
SL-2000 Maribor, Slovenia
phone:+ 386-222-849-00
fax:+ 386-225-194-82
Email: polonca.repic@guest.arnes.si
Demeter inspections in: Croatia, Slovenia.

Skal International India
No. 191, 1st Main Road
IND-Mahalaxmi Layout, Bangalore - 560 086, India. Phone:+ 91-80-23491928
Fax:+ 91-80-23491935
Email: skalindia@vsnl.com

Other Biodynamic Associations

Biodynamic Agriculture Australia Inc.
PO Box 54 AUS-NSW 2454 Bellingen
Australia. phone: +61-2-66-55-0566
fax:+ 61-2-66-55-0565
Email: poss@midcoast.com.au
Agrosofia
Krasnaja, 20, RUS-141500 Solnechnogorsk, Russia
phone: + 7 95 994 0397, fax: + 7 95 994 0397
Email: info@biodynamic.ru
internet: www.biodynamic.ru

Drustvo za biolosko-dinamicko
gospodarenje "DUGA" ,Tina Ujevica 39
HR-40000 Cakovec, Croatia
phone: + 385-40-38-40-52
fax: + 385-40-38-40-52
Email: bernarda.orehovec@ck.tel.hr
internet:www.demeter.hr

PRO BIO
Biologisch-Dynamische Sektion
Nemocnicni 53, P. O. Box 116
CZ-78701 Sumperk, Czech Republic
phone: 42-649-21-6609
fax: + 42-649-21-4586
Email: pro-bio@pro-bio.cz
internet: www.pro-bio.cz

Biodinamikus Kôzhasznú Egyesület
Eötvös u. 24, H-7090 Tamási, Hungaria
Kudraji, Inara Jansone, Aizputes pag.
LV-3456, Latvia
phone: + 37 134 486 12
fax:+ 37 165 171 79
Email: velgaj@navigator.lv

Fundatia Kaspar Hauser
Str. Gh. Lazar 16, RO-2700 Deva
Romania. phone: +40-254-218-225
fax:+40-254-235-436
Email: marius.gabor@lemninvest.ro

Contact in Greece
Gabriel Panagos, Rodopis Str. 8 A
GR-15234 Halandri, Athen Greece
phone: + 30 210 - 680 17 56 (morning)
fax: + 030 210 - 680 17 56

Biodynamic contact in Chile
Francisco Valenzuela Schulz
Parcela A 13, Fundo Alhué, Curacavi / Región Metropolitana, Chile
phone: +56 9 438 0771
Email: agriculturabiodin@yahoo.com

Asociacion de Agricultura Biodinamica - Peru
Pedro Otero, Corot 180, Lima 41, Peru
phone: + 51 14 35 27 88
Email: biodinamicaperu@yahoo.com

"Lebendige Erde" in Ukraine
Alexander Denisenko, Pickoba 3/6
79010 Lviv, Ukraine
phone: + 380 322 98 66 25
Email: tuag@is.lviv.ua

References

1. Dabbert, S., Häring, A. M. and Zanoli, R. 2004. *Organic Farming- Policies & Prospects*. Zed Books Ltd., London, UK. p. 169.
2. Kilcher, L., Huber, B. and Schmid, O. 2004. Standards and Regulations. In: *The World of Organic Agriculture-Statistics and Emerging Trends 2004* (eds. H. Willer and M. Yussefi). IFOAM, 27-43.

3. Vogl. C. R., Kilcher, L. and Schmidt, H. 2005. Are standards and regulations of organic farming moving away from small farmers' knowledge? *J. Sustainable Agric.* **26**(1): 5-26.
4. Vossenaar, R., Jha, V. and Wynen, E. 2004. Standards, certification and accreditation. In: *Trading Opportunities for Organic Products from Developing Countries* (Eds. Vossen René. and Wynen, Els). 48-56, UNCTAD, United Nations.
5. Schmid, O. 2002. Comparison of EU Regulations 2092/91, Codex Alimentarious Guide lines for Organically Produced Food 1999/2001, and IFOAM Basic Standards 2000. *In: International Hermonization and Equivalence on Organic agriculture. IFOAM Conference on Organic Guarantee Systems.* FAO and UNCTAD, February, 2002.
6. Riddle, J. and Coody, L. 2002. Comparison of the EU and US organic Regulations. *In: International Hermonization and Equivalence on Organic agriculture. IFOAM Conference on Organic Guarantee Systems.* FAO and UNCTAD, February, 2002.
7. Gaskell, M. 2000. Organic certification, farm production planning and marketing. Vegetable Research Information Centre, University of California, Davis, C, Publication 7247.
8. Rundgren, G. 2004. Organic certification statistics: 364 certification bodies in 57 countries. In: *Trading Opportunities for Organic Products from Developing Countries* (eds. Vossen René. and Wynen, Els). 47-54, UNCTAD, United Nations.
9. Bächi R. 2001. Lowering certification costs for small scale organic producers. *Paper presented at an UNCTAD Expert Meeting, Geneva,* July 2001.
10. IOAS (International Organic Accreditation Service) 2001. 'Accreditation update'. Newsletter No. 201, May 2001, Available from the internet at http://www.ifoam.org/ accredit/news0501final.pdf. 14 August 2001.
11. Bächi, R. 2003. Regulated organic agriculture system turns against farmers. In: *The Organic Standard.* Issue 21/January 2003 Grolind Ab, Hoje, Sweden: p. 18.
12. Zygmont, J. 2000. US organic fruit: export opportunities and competition in the international market. Paper presented at the *96th Annual Meeting and Trade Show of Washington Horticultural Association.* 6 December, 2000. Yakima, Washington, DC
13. Rundgren, G. 1999. The challenges of developing countries to establish an organic guarantee system. Paper presented at the *6th IFOAM Conference*, Florence, October.

GLOSSARY

Accreditation

The evaluation and formal recognition of a certification programme by an authoritative body.

Accredited programme

Means programme of accrediting inspection and certification agencies which have been accredited by the accreditation agency and which have agreed to comply with the accreditation contract.

Agribusiness

Includes all the activities that take place in the production, manufacturing, distribution, wholesale and retail sales of an agricultural commodity.

Alternative farming

Non-conventional, but not necessarily organic farming-although often used as a synonym for the latter.

Annual report

Means the report on operators, products and processors submitted annually to the accreditation agency by the accredited inspection and certification agencies.

Antioxidants

Antioxidants are compounds that help to inhibit many oxidation reactions caused by free radicals thereby preventing or delaying damage to the cells and tissues. Their mechanisms of action include :- (i) scavenging of reactive oxygen species and reactive nitrogen species (2) chelating metal ions prevent the generation of free radicals (3) decreasing the localized oxygen concentration thereby reducing molecular oxygen's oxidation potential (4) metabolizing lipid peroxides to non-radical products.

Audit, auditor, auditing body

See: inspection, inspector, inspection body.

Certification

A procedure by which a third party gives written assurance that a product, process or service is in conformity with certain standards.

Certification body

An organization performing certification. Sometimes referred to as the certifier or the certification agency. The certification body may use an existing standard or may set its own standard, perhaps based on an international and/or normative standard.

Certification label

A label or symbol indicating that compliance with specific standards has been verified. Use of the label is usually controlled by the standard-setting body.

Certification programme

A system of rules, procedures and management for carrying out certification. Sometimes referred to as a certification system. One certification body may execute several different certification programmes.

Codex Alimentarius

A set of standards on maximum chemical residues in food devised by a committee established by the Food and Agriculture Organisation (FAO) and the World Health Organisation (WHO).

Appeal

Shall be the process by which an inspection and certification agency can request reconsideration of a decision taken by the accreditation agency or an operator can request reconsideration of a decision by the certification agency.

Buffer zone

A clearly defined and identifiable boundary area bordering an organic production site that is established to limit application of, or contact with, prohibited substances from an adjacent area.

Certificate

Would mean a document issued by an accredited agency declaring that the operator is carrying out the activities or the stated products have been produced in accordance with the specified requirements in accordance with the particular standards.

Certificate of registration

Shall mean the document issued by the inspection and certification agency, declaring that the operator is licensed to use the certificate on specified products.

Certification

Shall be the procedure by which a written assurance is given by the certification agency that a clearly identified production or processing system has been methodically assessed and conforms to the specified requirements.

Consultancy

Shall mean the advisory service for organic operations, independent from inspection and certification procedures.

Consumer protection

Consumer protection laws are statutes governing sales and credit practices involving consumer goods. Such statutes prohibit and regulate deceptive or unscrupulous advertising and sale practices, product quality, credit financing and reporting, debt collection, leases, and other aspects of consumer transactions. The goal of consumer protection laws is to place consumers, who are average citizens engaging in business deals such as buying goods or borrowing money, on an even par with companies or citizens who regularly engage in business.

Contamination

Pollution of organic product or land; or contact with any material that would render the product unsuitable for organic certification.

Control body

Terms commonly used by the trade when referring to inspection and an inspection body.

CTC

It stands for crush, tear and curl. CTC machine consists of two toothed rollers which move at different speeds in opposite directions. The leaf moves continuously between the rollers. CTC machine produces smaller pieces than the orthodox rollers.

Conventional

Farming systems dependent on input of artificial fertilizers and/ or chemicals and pesticides or which are not in conformity with the basic standards of organic production.

Conversion

The process of changing an agricultural farm from conventional to organic farm. This is also called transition

Export subsidies

Special incentives provided by governments to encourage increased foreign sales. Subsidies, which are contingent on export performance, may take the form of cash payments, disposal of government stocks at below-market prices, subsidies financed by producers or processors as a result of government actions such as assessments, marketing subsidies, transportation and freight subsidies, and subsidies for commodities contingent on their incorporation in exported products.

Fermentation/ oxidisation of tea leaves

The reaction that tea leaves undergo when they have been cut and exposed to air. The extent of oxidisation determines whether the final tea is green, oolong or black.

Flavonoids

Flavonoids are a subclass of the polyphenols, which are characterized as containing two or more aromatic rings, each bearing at least one aromatic hydroxyl and connected with a carbon bridge. It is found predominantly in tea, have been associated with a wide range of biological activities, including antiangiogenic, antibacterial, anticarcinogenic, anti-inflammatory, antithrombobic, antiviral, hypocholesterolemic, hypoglycemic and vasodilatory effects.

Farm unit

An agricultural farm, area or production unit managed organically, by a farmer or a group of farmers.

Food additive

Food additive is an external permissible ingredient added to improve the keeping quality, consistency, colour and other physico chemical, sensory properties, wholesomeness and safety of food.

Green manure

Manure consisting of fresh green plant matter, which is ploughed in or turned into the soil for the purpose of soil improvement.

Group certification

Certification of an organized group of producers, processors and exporters with similar farming and production systems and which are in geographical proximity.

Hormone

A chemical messenger produced by endocrine glands and secreted directly into the blood stream to exert a specific effect on a distant part of the body.

Inputs restricted

Those items that are allowed in organic farming, in a restricted manner, after a careful assessment of contamination risk, natural imbalance and other factors arising out of their use. Farmers should consult the certifying agency.

Inspection

An on-site visit to verify that the performance of an operation is in accordance with specific standards of a certification programme.

Inspector

The person appointed to undertake the inspection, and may be an independent operator or an employee of the certifier.

Inspection body

The body performing the inspection part of certification. Where a certification body performs its own inspections, the certification body is also the inspection body.

Integrated farming

A farming system that aims to reduce the environmental impact of agriculture while pursuing economic goals. Standards are voluntary. The farming system is not legally certified.

Licence

A document issued under the rules of a certification programme, by which a certification body grants a person or body the right to use certificates or certification labels for its products, processes or services in accordance with the rules of the relevant certification programme.

Livestock

Shall mean any domestic or domesticated animal including bovine (including buffalo and bison), bovine, porcine, caprine, equine, poultry and bees raised for food or in the production of food. The products obtained by hunting or fishing of wild animals shall not be considered as part of this definition.

Marketing

Means holding for sale or displaying for sale, offering for sale, selling, delivering or placing on the market in any other form.

National accreditation body

Shall be the agency set up by the steering committee for National Programme for Organic Production for accrediting inspection and certification agencies.

Non-renewable resources

A natural resource, such as fossil fuels that have a finite stock and cannot be renewed.

Orange pekoe

Referring to size of leaf, not quality or flavor, this term indicates a larger-size grade of whole leaf teas.

Organic

Refers to a particular farming system as described in these standards and not to the term used in chemistry.

Organic farming

A set of agricultural practices in which no artificial fertilizers, hormones, pesticides are used. Organic farming is increasingly popular in developed countries as consumers become more concerned over the potential health hazards of chemical residues in food. More detailed definition provided in Chapter 1.

Orthodox

This is the traditional method of processing. Prepared using a technique which leads to larger leaf styles mirroring hand-produced teas.

Pan fired

Tea that is steamed and then agitated in an iron wok over a fire.

Parallel production

Shall mean any production where the same unit is growing, breeding, handling or processing the same products both in a certified organic and a non-certified organic system. Similarly a situation with "organic" and "in conversion" production of the same product is also parallel production.

Pekoe

Whole leaf black tea produced by a medium plucking of the second leaf on the tea bush (The word Pekoe comes from the Chinese meaning 'white hair' and was originally applied to the early tea pluckings, due to the white down on the backs of the young leaf tea).

Plant protection product

Shall mean any substance intended for preventing, destroying, attracting, repelling, or controlling any pest or disease including unwanted species of plants or animals during the production, storage, transport, distribution and processing of food, agricultural commodities, or animal feeds.

Preparation

Shall mean the operations of slaughtering, processing, preserving and packaging of agricultural and animal products and also alterations made to the labeling concerning the presentation of the organic production method.

Price premium

The extra money received by farmers or paid by consumers for a quality/premium product.

Standards

Documented agreements containing technical specifications or other precise criteria to be used consistently as rules, guidelines or definitions, to ensure that materials, products, processes and services are fit for their purpose. Standards include environmental standards; organic standards; labour standards; social standards; and normative standards.

Environmental standards are standards for materials, products and production processes to ensure that negative impacts on the environment are minimal or kept within certain limits.

Organic standards are standards for production and processing of organic food products. Labour standards are standards for working conditions to ensure workers rights are respected.

Social standards can be used to mean labour standards, but can also include standards on other social aspects of organizations and production facilities, such as the relation with neighbouring communities.

Normative standards are generic (general or non-specific) standards or guidelines to be used as a framework by local standard-setting or certification bodies when formulating a specific standard for their certification programme. Normative standards are also referred to as Standards for Standards, e.g. the IFOAM Basic Standards and FAO/WHO Codex Alimentarius guidelines.

Sustainable agriculture

Method of farming that does not degrade the productive capacity of the land.

Organisations with interest in organic farming

International Organisations

IFOAM
c/o Oekozentrum Imsbach
66636 Tholey-Theley
Germany
Tel: + 49 6853 919890
Fax: + 49 6853 919899
http://www.ifoam.org
headoffice@ifoam.org

Food and Agriculture Organisation of the United Nations (FAO)
Viale delle Terme di Caracalla
00100 Rome
Italy
Tel: + 39 06 5705 1
Fax: + 39 06 5705 3152
http://www.fao.org
fao-hq@fao.org

Pesticides Action Network (PAN) Europe
Eurolink Centre
49 Effra Road
London SW21BZ
UK
Tel: + 44 20 7274 8895
Fax: + 44 20 7274 9084
http://www.pan-europe.net
coordinator@pan-europe.de

Organic Trade Association
PO Box 547
Greenfield, MA 01302
USA
Tel: 1413 774 7511
Fax: 1 413 774 6432
http://www.ota.com
info@ota.com

International Trade Centre UNCTAD/WTO (ITC)
Palais des Nations 1211
Geneva 10, Switzerland
Tel: + 41 22 730 01 11, Fax: + 41 22 733 44 39
http://www.intracen.org
itcreg@intracen.org

Research Organisations

Forschungsring fur Biologisch-Dynamische Wirtschaftsweise
Brandschneise 1
64295 Darmstadt, Germany
Tel: + 49 6155 84123, Fax: + 49 6155 846911
http://www.forschungsring.de
Immo.Luenzer@Forschungsring.de

Forschungsinstitut fur biologischen Landbau (FiBL)
Ackerstrasse
5070Frick, Switzerland
Tel: + 41 62 865 72 72, Fax: + 41 62 865 72 73
http://www.fibl.ch
admin@fibl.ch
Web-Project: http://www.organic-europe.net

Louis Bolk Institute
Hoofdstraat 24
3972 LA Driebergen, Netherlands
Tel: + 31343 52 38 60, Fax: + 31343 5156 11
http://www.louisbolk.nl
info@louisbolk.nl

The Rodale Institute
611 Siegfriedale Road
Kutztown, PA 19530-9320, USA
Tel: + 16106831400, Fax: + 1610 683 8548
http://www.rodaleinstitute.org
info@rodaleinst.org

Ludwig-Boltzmann-Institute for Biological Agriculture and Applied Ecology
Rinnbockstrasse 15
1110 Vienna, Austria
Tel: + 431 7951497943, Fax :+ 43 1 795147393
http://www.organic-research.com/research/Projects/DOCEA/inst4.asp

HDRA
Ryton Organic Gardens, Coventry
Warwickshire CVS 3LG, Great Britain
Tel:+ 44 24 7630 3517, Fax: + 44 24 7663 9229
http://www.hdra.org.uk
enquiry@hdra.org.uk

Elm Farm Research Centre
Hamstead Marshall
Newbury RG20 OHR, Great Britain
Tel: + 44 1488 658298, Fax:+ 44 1488 658503
http://www.efrc.com
elmfarm@efrc.com

The Norwegian Centre for Ecological Agriculture
6630 Ingvoll, Norway
Tel: + 47 7153 20 00, Fax: + 47 71 53 20 01
http://www.norsok.no
norsok@norsok.no

Danish Research Centre for Organic Farming (DARCOF)
Foulum, P.O. Box 50, 8830Tjele
Denmark
Tel: + 45 89 99 10 35, Fax: + 45 89 99 16 75
http://www.darcof.dk
foejo@agrsci.dk

Centre for Sustainable Agriculture (CUL)
Box 7047, 750 07 Uppsala
Sweden
Tel:+ 46 18 67 10 00, Fax:+ 46 18 67 35 71
http://www.cul.slu.se
jessica.alm@cul.slu.se

Institut Technique de 1'Agriculture Biologique (ITAB)
149, rue de Bercy
75595 Paris cedex 12, France
Tel: + 33 1 40 04 50 64, Fax:+ 33 140 04 50 66
http://www.itab.asso.fr
itab@itab.asso.fr

Henry A. Wallace Institute for Alternative Agriculture
9200, Edmonston Road, Suite 117 Greenbelt, Maryland, MD 20770-1551 USA http://www.hawiaa.org

Some Internet Sites on Organic Agriculture and Organic Market

http://www.agric.wa.gov.au/agency/Pubns/farmnote/2000/f13700.htm
http://www.agriculture-industry-india.com/agro-programme-schemes/fertiliser1.html
http://www.alphaomegafood.com/olive_oil.htm
http://www.apeda.com/organic.htm
http://www.aryanint.com/aryspice.asp
http://www.biofach.de/main/page.html
http://www.codexalimentarius.net/web/index_en.jsp
http://www.djmiles.co.uk/
http://www.dtrdc.org/
http://www.eat-japan.com/shopping/clearspring/
http://www.ers.usda.gov/
http://www.etagriculture.com/mar_april2002/trade.html
http://www.fao.org/organicag/
http://www.fantasticfoods.com/sidedishes.php
http://www.fas.usda.gov/agx/organics/news2000.html
http://www.fas.usda.gov/info/agexporter/2000/June/germany.htm
http://www.fibl.ch/english/fibl/annual-report-2002-international-indocert.pdf
http://www.fibl.ch/publikationen/pdfs/verzeichnisse/handbook-coffee.pdf
http://www.fordfound.org/publications/recent_articles/docs/Solutions_40-45.pdf
http://www.geocities.com/RainForest/7813/0831_cop.htm
http://www.ginschis.com/green_t.html
http://www.goodnessdirect.co.uk/cgi-local/detail/326737.html
http://www.greens.org/s-r/28/28-10.html
http://www.hangaofood.com/english/organic%20in%20the%20world/Poland.cfm
http://www.highlandorganics.co.uk/pastaproducts.htm
http://www.ifoam.org
http://www.maff.go.jp
http://www.ileia.org/
http://www.indianorganic.info/int/e_prog.htm#top
http://www.indianspices.com/html/s0650org.htm
http://www.indianspicesboard.com/html/s0650org.htm
http://www.indocert.org
http://www.intracen.org/mds/sectors/organic/welcome.htm
http://www.ipo.org.za/news/archive/oct02.htm
http://www.jinjapan.org/trends/article/021022sci_r.html

http://www.jinjapan.org/trends/article/021022sci_r.html

http://www.konarainforestcoffee.com/history.html

http://www.lightparty.com/Health/OrganicAgricultureWW.html

http://www.linksorganic.com/uk/links.asp?Activity=Broker

http://www.marketag.com/ma/news/archive/v31/nuts1.stm

http://www.obcebdbh.be/import_en/info-center/beci-reviews/2002/ES-biofood/part6_en.html

http://www.obcebdbh.be/import_en/info-center/beci-reviews/2002/ES-biofood/part5_en.html

http://www.organicconsumers.org/Organic/nitrofen060502.cfm

http://www.organicdelivery.co.uk/

http://www.organic-europe.net/country_reports/france/default.asp

http://www.organic-research.com

http://www.organicts.com/services/tradeline.html

http://www.ota.com

http://www.pdsorganicspice.com/about.htm

http://www.purefood.org/Organic/spicesgood.cfm

http://www.soel.de/oekolandbau/weltweit_reports.html

http://www.squ.edu.om/lan/Science2051/atech/perma.html

http://www.stassengroup.com/inpages/co_busi.htm

http://www.statpub.com/stat/open/0jkrfwl.html

http://www.tierramerica.net/2001/1118/iarticulo.shtml

http://www.tradenetsl.lk/info/cashew.html

http://www.varanashi.com

Index